Inclusive Computing Education in the Secondary School

Underpinned by pedagogical practices and theories of what works in teaching computing, this book gives existing and new teachers ideas to enable them to plan an inclusive curriculum for the secondary school computing classroom.

Computing is one of the fastest-developing subjects in the curriculum, and computing teachers will always be updating their subject knowledge and pedagogical approaches. Each chapter explores a specific aspect of inclusion and potential barriers faced by students and is designed to challenge teachers to think about their own practice and curriculum design. Themes include the influence of classroom environments, bias in the use of data, collaborative learning, building cultural capital, and racism within AI applications. The book is also laced with practical ideas to develop teaching shared by a wealth of experienced practitioners, researchers and industry professionals.

Written with consideration for the National Curriculum for Computing, this valuable text will give trainee teachers, recently qualified teachers, and experienced teachers the confidence and knowledge they need to successfully deliver an inclusive computing curriculum in the classroom.

Louise Hayes is a Senior Lecturer in Initial Teacher Education at Manchester Metropolitan University, UK.

Eleanor Overland is the Director of Quality Assurance for Initial Teacher Education at Manchester Metropolitan University, UK.

Inclusive Computing Education in the Secondary School

Linking Theory and Practice

Edited by Louise Hayes and Eleanor Overland

LONDON AND NEW YORK

Designed cover image: © Getty Images

First published 2024
by Routledge
4 Park Square, Milton Park, Abingdon, Oxon OX14 4RN

and by Routledge
605 Third Avenue, New York, NY 10158

Routledge is an imprint of the Taylor & Francis Group, an informa business

© 2024 selection and editorial matter, Louise Hayes and Eleanor Overland; individual chapters, the contributors

The right of Louise Hayes and Eleanor Overland to be identified as the authors of the editorial material, and of the authors for their individual chapters, has been asserted in accordance with sections 77 and 78 of the Copyright, Designs and Patents Act 1988.

All rights reserved. No part of this book may be reprinted or reproduced or utilised in any form or by any electronic, mechanical, or other means, now known or hereafter invented, including photocopying and recording, or in any information storage or retrieval system, without permission in writing from the publishers.

Trademark notice: Product or corporate names may be trademarks or registered trademarks, and are used only for identification and explanation without intent to infringe.

British Library Cataloguing-in-Publication Data
A catalogue record for this book is available from the British Library

ISBN: 978-1-032-04538-2 (hbk)
ISBN: 978-1-032-04540-5 (pbk)
ISBN: 978-1-003-19368-5 (ebk)

DOI: 10.4324/9781003193685

Typeset in Melior
by KnowledgeWorks Global Ltd.

Contents

About the editors   vii
About the contributors   viii
Special mention   xi
Foreword by Beverly Clarke   xii

How to use this book   1
Emma Merva and Eleanor Overland

1. The developing curriculum: From ICT to computing   7
   Eleanor Overland

2. The development of artificial intelligence in computing education: Thinking betwixt and between – reinvigorating Papert's im/possibilities of computing   19
   Amanda Banks Gatenby

3. Keeping it real: Helping learners navigate the concrete and abstract   34
   Mick Chesterman

4. AI is racist   45
   Richard A. Dunk

5. Using data to ensure an engaging and inclusive computing curriculum   60
   Matthew Thorpe

6. Opting out: Why are pupils choosing not to study computing?   75
   Cathy Lewin and Eleanor Overland

7. Gender differences in computing classrooms: Practices to develop inclusive learning spaces   84
   Louise Hayes

8 Design and project approaches in computing education 95
   Mick Chesterman

9 Industry perspectives 108
   Louise Hayes and Eleanor Overland

   Index 120

About the editors

Louise Hayes is a Senior Lecturer in Education at Manchester Metropolitan University, where she is passionate about the subject of computing and understands how the subject can help to raise the aspirations of young people and help them in their future careers. She is a Director of an Education Company and a Professional Development Lead for STEM Education. She started work straight from "A" Levels, where she changed her career from BT plc into secondary school teaching; this was supported by government funding to enable her to further her studies. In her doctorate studies, she has written about the issues in education from a gender perspective, the influences of computing classroom environments on girls, and the impact of this on subject uptake and career trajectories of women in tech. Travel has played a large part in her career, where she has led a number of school enrichment visits, and more recently, as a member of the iFiP TC3 group, she has delivered tech workshops to over 70+ teachers in India and supported outreach computing teaching in Malawi.

Eleanor (Ellie) Overland is the Director of Quality Assurance for Initial Teacher Education at Manchester Metropolitan University (MMU). She oversees quality assurance across primary and secondary teacher training at MMU and, prior to this, led the secondary PGCE programme. Ellie is a specialist in computing education in secondary schools. She launched the Computing PGCE at MMU in 2013 and has trained many wonderful computing teachers. She also provides CPD for highly experienced staff in developing computing subject knowledge and pedagogies for their classrooms. She is also interested in online learning opportunities for teachers and is the lead educator on two MOOCs on the Futurelearn platform, one for those thinking of teaching and another for early-career teachers. So far, those courses have attracted over 20,000 learners. She is also passionate about research and the continuing development of computing education curriculum and teaching practice. She is currently completing her doctoral thesis, which focussed on the implementation of the computing curriculum in contrasting schools.

About the contributors

Mick Chesterman teaches at the Manchester Metropolitan Faculty of Health and Education primarily on foundational and project-based units. His PhD studies involve families exploring the cultural and ecological issues of coding platform games together. He has a history of teaching media-making and web-creation skills to facilitate positive change for social groups. In recent years, he has run several outreach workshops in local communities and schools linked to the university and in his home area of the Calder Valley. These workshops have a focus on creative coding and the use of physical materials. Favourites include music making using code, physical computing using crumbles, code bugs and microbits and game making using web technologies. He has recently co-founded a making and repairing workshop called Todmorden Makery, which works with adults, young people and families to repair objects and transform old technology into art projects.

Beverly Clarke is a TechWomen100 award winner (2022). She is a former national manager for Computing at School (CAS) – the teacher-facing arm of BCS – The Chartered Institute for IT. In 2022, she was also a Computer Weekly longlist nominee for most influential women in UK Tech. She currently works as an author, education consultant, coach, mentor, ambassador and trustee. She saw her first computer at the age of 12. This sparked an interest in technology for her and an awareness of how the world was divided place through technology and the opportunities available for those with access to the digital world. Her journey in Computing and Technology education started in the corporate IT sector and then led to retraining to work in education and teaching. She is also a former computing teacher and the Head of Computing department. Passionate about technology and the impact it is making upon all of our lives, she is the author of two published books "Computer Science Teacher: Insight Into the Computing Classroom", aimed at attracting new entrants into the teaching profession and "The Digital Adventures of Ava and Chip", a children's book series with a technology focus. Her work has national and international coverage.

Richard A. Dunk is a Senior Lecturer in Education at Manchester Metropolitan University. He started programming computers at the age of eight, hacking the QBasic game *Gorillas* to make sure that his younger brother could never win. This programming experience served him well in obtaining a Master's degree in theoretical physics, which then led to nine years of experience teaching science, mathematics and computing in high schools and FE institutions. Now working in teacher education and research, his current research interests revolve around post-qualitative methodologies, arts-based methods, and the use of advanced computational tools in socio-material analysis.

Amanda Banks Gatenby is a Lecturer in Digital Technologies and Communication in Education at the University of Manchester. With early interests in the process of learning developing from her attempts to learn and teach the biomechanics and psychology of figure skating, she abandoned GCSE computing after three months in year ten and didn't touch a PC again until a few years later, when a Compaq 486 appeared in the office where she worked. As no one else wanted to learn how to use this strange machine, this began a journey from novice user to student of computer science, Software Engineer at Marconi Communications, Project Manager at a subsidiary of RM, Teacher of ICT, learning technologist at a School of Computer Science and PhD study of how young people learn computational perspectives through working with Raspberry Pi Technologies. Amanda now works with educators in schools, higher education and informal settings, teaching a range of subjects, including A.I. perspectives on learning, learning experience design, learning analytics and other aspects of digital education.

Cathy Lewin is a Professor of Education at Manchester Metropolitan University. As a teenager, she received a Sinclair ZX80 kit, which she assembled and soldered together before experimenting with coding. She went on to gain a BSc in mathematics with computer science and a PhD in educational technology. She has conducted research on the use of technology to support teaching and learning in school classrooms for over 25 years. This includes teachers' digital pedagogies, young people's uses of technology for education and personal interests, the impact of social media on teaching and learning, and the blurring boundaries between formal and informal learning. One of her current interests relates to girls and computer science career pathways and how this is affected by qualification option constraints and the need to feel a sense of belonging. She draws on socio-cultural theoretical frameworks, including activity theory and figured worlds.

Emma Louise Merva is a highly experienced national and international educational professional. She has worked across the world in both an advisory and inspection capacity with governments and educational providers across the United Kingdom and Middle East region. She has taught, examined and inspected Computer Science and Business Education for over 24 years. Currently, she is Regional Director for one of the largest Multi-Academy Trusts (MAT) operating across England. Prior to this, she was the founding Executive Principal and set up two outstanding free schools in the Northwest. She has sat on several strategic boards across the Northwest region, having been in senior strategic roles since 1997. She facilitates and coaches on several postgraduate leadership masters' modules in addition to her day-to-day work. She holds several postgraduate degrees. Including the NASEN award, she has a postgraduate certificate in SEND, a postgraduate degree in Early Childhood Studies, a postgraduate degree in HR, a TEFL certificate and a Master's degree with a focus on SEND. She is currently on the thesis stage of an Education Doctorate (EdD) and has spent the last three years gaining experience at the doctoral level in the use of ethnographic and case study research methods. Her thesis focus is on the history and leadership of virtual learning in compulsory schooling (3–16).

Matt Thorpe is a Senior Lecturer in Education at Manchester Metropolitan University. He works with both primary and secondary computing students on the Postgraduate Certificate in Education. He has a BSc (Hons) in multimedia and internet technologies and a master's degree in education. He is currently studying for his PhD, investigating student and staff cultures of data within higher education. His research interests more broadly relate to digital education and the impact of datafied practice within education. He worked for eight years within further education, leading the delivery and development of A-level and vocational ICT and iMedia programmes. Matt has also worked as a learning technologist within higher education, supporting academic staff to more effectively embed technology into their practice.

Special mention

Guy Hayes, MSc International Business Masters student at Leeds University, thank you for your help with proofreading the final chapters.

Emily Overland, age 13, who chose the cover image and design for the book – a great choice:).

Foreword

In my life, there have been two experiences that have profoundly influenced my professional career. The first was seeing a computer for the first time at the age of 12 whilst visiting the United States. The other was standing in a classroom in 2002, preparing to deliver my first early evening lecture to a group of A-level students re-sitting their exams alongside adult learners who were studying for an Information and Communication Technology (ICT) qualification.

In my first experience, I observed other children playing computer games and interacting with the world in a way of which I was unaware – this was not something that was available to me growing up in a less developed country, Guyana. I also remember being totally curious and aware that there was a very different world of opportunities and possibilities. I simply wanted to know more!

The second experience presented me with my first real appreciation of the challenges of teaching topics in this subject. Specifically – how could I impart the required knowledge and skills in ways with which the learners would connect? I was also aware of how the lives of each and every learner could be ultimately transformed through gaining knowledge and skills in ICT. In particular, the opportunities that could potentially emerge for each person as a result of understanding the subject matter.

Fast forward to 2014, and there was a change to the National Curriculum for England, which led to the discrete teaching of computing and computer science in schools, with emphasis on areas such as computational thinking, creativity, programming and understanding of the changing world. In fact, curriculum changes have affected not only the English national curriculum but global curricula. These changes have raised many questions and theories about the best way to address each of these topic areas. For example, the topic of computational thinking has seen a need to ensure that concepts and approaches utilised when teaching computing have parallels drawn with successful STEM[1] subjects such as engineering. There is also a need for improved pedagogy in our subject to ensure that key concepts are fully understood.

The authors of this book take the reader on a journey through crucial considerations in teaching this subject, looking at practical and theoretical aspects gained from their own practice, from industry and also from research conducted by leading institutions such as the Royal Society.

The examples used give the reader a variety of valuable ideas to foster a more inclusive computing classroom, with examples such as encouraging computing outside the classroom and the curriculum to develop wide-scale acceptance of computing. The resulting societal impacts of this will be an economy with increased productivity and citizens who are able to operate fully in the digital world. This is particularly important within areas such as culturally responsive pedagogy, which contributes to ensuring that learners understand their cultural place in the world, along with the benefits that access to technology can bring to them and their communities. This ultimately leads to greater digital literacy awareness, increased participation in world affairs, and greater social mobility.

Thinking back to my early experiences of teaching this subject and the learning environments in which computing is generally delivered, it is great to see this book explore the use of data to support teaching and learning, alongside the use of classroom pedagogies. This is particularly pertinent with increasing educational products utilising artificial intelligence and data to deliver personalised learning solutions.

Furthermore, we need to address the issue of low numbers of women and minority groups choosing to study this subject in both schools and colleges, along with the small numbers of these groups pursuing technology careers. I'm part of the 0.7% of ethnic minority (black) women in a tech career.[2] It is heartwarming to see this book address initiatives to encourage female uptake in computing and address some of the stereotypes that lead to a poor perception of technology careers for women. The book also explores other aspects of inclusion within computing, including race and socio-economic standing.

Therefore, for me, writing this foreword is a great reflection on the progress and positive developments starting to be made in our subject. Having taught computing, managed and led national projects, and contributed to international projects, there is still some way to go to move beyond computing being seen as a challenging subject and the preserve of the few to being accepted as a necessary life skill for the many. Looking ahead, I am provided with much hope that computing can be fully adopted and embraced by all.

As a national leader in computing education, it gives me much pride and pleasure to see a part of this book dedicated to supporting professional development within computing education, a topic close to my heart. Being a part of a computing professional development community has helped me and countless others in their careers. It is through this coming together with a common theme of computing education to improve our practices that we learn and improve. As a result, the learners in our care benefit by gaining the necessary knowledge and skills to function effectively, find their place in the digital world, and make progress to reduce the digital divide that affects so many.

My hope is that this book – *Inclusive Computing Lessons for Secondary School Education*, elevates the discussion around computing education while also showing a rounded view of what is required to deliver an inclusive classroom in which all learners have their individual needs met and the best opportunities made available to them through a high-quality computing education.

Beverly Clarke
Author and National Leader in Computing Education

Notes

1 STEM – Science, Technology, Engineering, and Mathematics.
2 Diversity and inclusion | BCS.

How to use this book

Emma Merva and Eleanor Overland

Introduction

As Beverley so honestly explains in her foreword, exposure to computing can really open up life chances for young people. This book explores some of the current issues around inclusivity in computing. The book also outlines theory and practice to support computing teachers to develop inclusive practice in their own classrooms. As you explore, you'll find the situation and the underpinning causes are complex and the authors cannot pertain to have all the answers. Having started to read this book yourself, you clearly have an interest into how computing can become more inclusive. These chapters will provide space to think around the issues and theory on which to build your practice and practical ideas you can try in your own teaching. You may choose to read this book from cover to cover, although you may choose to dip into certain chapters you find most relevant. Either way will work. Here is a short summary of each chapter to help guide you through your approach to using this book.

The first chapter starts at the beginning of the story. How did computing as a subject develop and why do we have some of the challenges around inclusion that we can currently see in classrooms. It challenges the reader to consider the actual curriculum design. Are we teaching the right things, in the right order? Does the curriculum design in your school unintentionally "exclude" groups of pupils. Could a curriculum rethink lead to more inclusive practice?

The second chapter draws on the seminal theories of Seymour Papert but brings them to life through a relatively new area of education, Artificial Intelligence (AI). AI is one of the largest growth areas in computing and so it is important educators have an understanding and are able to adapt their practice accordingly. Computing is one of the fastest developing subjects in the curriculum and computing teachers will always be updating their subject knowledge and pedagogical approaches.

Chapter three takes a practical approach to exploring computational thinking. Some concepts for pupils can be difficult to grasp, especially for pupils who may have special educational needs. In this chapter a range of pedagogical approaches are explored. These can be taken by the reader and used within their teaching to support pupils to understand concepts and so know and remember more.

Chapter four takes a fascinating look at computing applications already available in the industry and how these themselves can be less than inclusive. The chapter reinforces the message of the importance of a diverse computing workforce to try and ensure such prejudices can be addressed at the developer stage. The chapter also explores approaches that educators can take to address some of these.

Chapter five follows on closely from the challenges outlined in the previous chapter but considers specifically about data and how this is used within schools. The chapter explores some of the software and apps used by teachers, where there may be pitfalls and how some of these can be harnessed to support inclusive teaching environments.

Despite the best efforts of educators, there are still many disparities between pupils that select to study computer science and those that do not. In chapter six, levels of representation are explored, along with reasons pupils give for not electing to study computer science. The chapter concludes with some practical suggestions to try in school classrooms.

Chapter seven focusses specifically on the gender gap that exists within computing education. The chapter explores some of the statistics alongside some of the practical changes teachers can make in their classroom. These ideas include lesson planning and pedagogical approaches through to designs of computing spaces.

Chapter eight explores project-based learning, which some may view as a less formalised approach to delivering computing education. With more pupils being taught away from formal classroom environments (such as home-school) and more schools trying to offer meaningful extra-curricular opportunities, project-based learning may be an option. The chapter also explores how this approach could be used more regularly in the classroom as an integral part of the curriculum.

Chapter nine brings the book to a conclusion by speaking with colleagues in the computing industry and in research. The chapter explores views as to what teachers should or could be doing to make the industry more inclusive. The chapter also emphasises the scope of the challenge, but also the strength of feeling amongst all levels in the industry, that it is essential we get this right and ensure an inclusive start for pupils in the world of computing.

Becoming a computing teacher: Guidance for trainee and early career teachers

For new teachers, prior to reading the chapters you may find some additional information around computing education useful. In this section, practitioner, Emma Merva, offers some crucial information, advice and tips for becoming an inclusive computing teacher.

Computer science is an exciting and appealing subject area, which can stimulate all students. The National Curriculum for Computing in secondary education is enthralling covering all aspects of technology. As a computer science teacher, you can light up a student's day, give them the motivation they need to succeed, show them kindness and encouragement so that they become the best version of themselves. It is our responsibility as qualified professionals to understand the individual requirements of our students, guiding them, motivating them and giving them the educational experiences that they require to develop computing careers in occupations such as future app designers, computer programmers, gamers or AI robotic engineers, just to mention a few exciting vocations that the world of computer science can lead them into!

Classifying students correctly using assessment and attainment data to understand where each student's starting point and current progress will help you to ensure that their needs are sufficiently met. You can also identify students who are potentially underperforming. This will help you to target your teaching and identify what is having impact. It can support you to keep students highly engaged in the subject area, feel comfortable with the learning experience, develop their passion for the subject, and ultimately believe that this subject is for them. In short, the students must believe in you as a teacher, they need to buy into the subject area, believe that computer science is for them and that it has everyday relevance which impacts on their lives.

The Ofsted framework alongside the Computing programmes of study and Teacher Standards are important documents in determining how we teach each subject area (DfE, 2011). For example, the teaching in each lesson will need a justification of why Computing is being delivered in a particular way or an understanding of the suitability of the chosen rationale.

Each teacher in the subject being delivered in school needs to understand the 3 I's the Intent (what is being taught), the Implementation (how is it being taught) and the Impact (are the students progressing). The 3 I's are the foundations of the current Ofsted framework, which support schools in determining pedagogy. Early Careers Teachers should have a working knowledge, for instance, of the basic understanding of the psychology of learning and development, scaffolding and modelling.

Teaching inclusively requires you to be able to use the behaviour management systems and policies of the school where you are working. In a computer science classroom, the practitioner can soon master the art of behaviour management in a simple yet effective way through the use of numbering the computing spaces and allocating students to numbered workstations with thought about which students will work well together. You may also have classroom management software installed. Not only can this support with behaviour management techniques, it can be particularly useful to support pupils with special educational needs or disabilities (SEND). Modelling activities on their screens can help scaffold learning, which can sometimes be difficult for students to follow from one central screen.

Practical advice for creating and inclusive classroom

Inspirational role models: If you can see it, you can be it! Ensure all of your resources and classroom displays have pictorial representation of everyone in computing careers and examples of terrific work students have produced during your lessons! Celebrate all work through classroom displays, shine a light on great work from the last lesson using the share screen function in Office 365, the visualiser or through a short student led show and tell starter activity of fantastic computer science work from the last lesson focusing on what they learnt last term, last week and last lesson with the threads of subject knowledge being very clear to all students.

Cognitive load theory and memory: Ensure that resources are appropriately sequenced, and learning should develop over time to allow all children or young people to independently interact with the learning materials. For example, ask students to create a timeline of their own learning and subject knowledge advancement at the end of each unit or half term as a homework project which also acts as a knowledge organiser. Chapter 1 provides more information around designing and sequencing an inclusive computing curriculum.

Question everyone!: Teacher led questioning should focus on everyone engaging and participating. Useful strategies include no hands up, using simple techniques to engage all young people in your classroom, such as lollipop pop sticks which can be numbered and the teacher requests a number to answer a question. Other methods and questioning techniques, for example, include numbers given to students and mini white board interaction. Through planning and adapting your teaching you ensure that your practice is inclusive of all students in the classroom. The NCCE have produced some guides for different pedagogies for teaching computing (https://teachcomputing.org/pedagogy).

Have strong subject knowledge: Ensure that your digital skills are up-to-date and that you can use all the required software tools to teach online, for example, with online platforms and relevant apps to support pedagogical practice. Being technologically fluent is similar to being fluent in any language, this means you can work and operate the software across many platforms and you can enhance the learning experience of the students you teach. You may wish to consider updating your skills with the National Centre for Computing Education courses, such as the computer science Accelerator, and this gives you a recognised qualification to indicate you understand the subject at a proficient level.

Learn from other subject areas: Some of the challenges faced by computing teachers are quite different in other subject areas. Other subjects may not have the same gender imbalances, or they may take very different approaches to teaching aspects of the curriculum such as numeracy skills. Take opportunities to observe other colleagues. This will support you to continually improve and adapt your practice.

Keep up to date in computing: The National Curriculum Programme of Study for Computing states that students should be able to function in a world of work as a competent digital operator. If we really consider that many of the careers and jobs that our students will do haven't even been thought about yet! In the 1990's or early 2000's being a social media influencer was a distant dream. Skills needed for becoming a popular and successful social media influencer and content creator include digital editing, creating video content on mobile devices and basic programming understanding how social media sites use algorithms to enhance your presence and views on the platform hadn't even been thought about. All students, regardless of their needs, have something to offer the world of Computing. It's an inclusive employment area where it doesn't matter when and where you work. For example, on Tik Tok and Instagram there are many visually impaired influencers and content creators working from home or the coffee shop setting up their own businesses and changing social norms. Many students who do not fit into the traditional notion of academic or vocational education may often make a substantial contribution to the digital or virtual world. Opportunities for students experiencing education today will be far more linked into digital platforms and virtual learning spaces. You only have to visit the Expo 2020 and BETT as a virtual participant to see and understand that the world will continue to develop and transform as countries collaborate and technology changes how we work (Expo, 2020).

Develop culturally relevant computing lessons: Many chapters of this book explore underrepresented groups and how to ensure they are included within your teaching. This takes thought and careful planning. It also requires levels of empathy and sensitivity to allow teachers to look at computing lessons from different perspectives. The Raspberry Pi Foundation have been working with computing teachers to explore some approaches to develop culturally relevant teaching in computing. Insights into their research and a guide for teachers can be found on the Raspberry Pi blog (Delivering a culturally relevant computing curriculum: new guide for teachers – Raspberry Pi Foundation).

Building a network

In many schools computing departments are small. They may also be constrained by the number of specialist teachers they have been able to recruit. Liaising beyond your school is sure way to build your own experiences and develop your ideas. Many local authorities or multi academy trusts have their own subject specialist networks. If there is not one in your setting it may be the right time to start one!

In addition, Computing at School (CAS) have many local community groups welcoming specialist and non-specialists to meet and discuss computing. The topics can vary but they are driven by the teachers that attend. Sessions can include CPD, sharing of teaching resources, talks from local employers and practical hands on sessions. You can find your local hub on the CAS website. You will need to join first, it is all free (www.computingatschool.org.uk/community). CAS also run "CAS Chat" on Twitter every Tuesday evening during term time.

STEM ambassadors are another good way to make contact beyond your school. STEM ambassadors support encouraging young people to follow careers in STEM subjects. They are particularly mindful of underrepresented groups and will do all they can to support inclusive work in the subject.

Arranging trips to local business and organisations can really bring computing alive for your pupils. Some pupils, particularly those from disadvantaged backgrounds, may really benefit from visits to outside organisations. There may be limited budgets to consider, but even a small scale local visit can inspire pupils. Additionally Bletchely Park, Milton Keynes demonstrates a number of stories of intelligence and code breaking in WWII and is well worth a visit.

Classrooms, pedagogy, and teaching are forever evolving, becoming more fluid and flexible so that the teacher can adapt and adopt to society's fast paced technological developments. The Computing Programme of Study states students should be able to operate and compete "… at a level suitable for the future workplace and as active participants in a digital world" (DfE, 2013). In theory, the use of technology actually has the power to increase access and inclusion for all. This book explores where that is not yet happening, and some pupils do not achieve within computing as much as we would hope.

The time in the classroom for the student is short, making each lesson count. Each lesson is an opportunity to ignite the passion and love of computing that could enable these students to become the future technological innovators that may change the world for the better. There is support available and many educators are working towards the same goal of making computing a more inclusive subject. We hope you are able to take key messages from this book to apply to your own teaching, and can build your network to develop practice with others. Increasing diversity and representation within the subject is complex but computing teachers do have the opportunity to make a difference.

References

DfE. (2013). National curriculum in England. Computing programmes of study: Key stages 3 and 4 National curriculum in England: computing programmes of study - GOV.UK (www.gov.uk).

DfE. (2011). Teachers' Standards - GOV.UK (www.gov.uk).

Expo (2020) Connecting minds and creating the future www.expo2020dubai.com

The developing curriculum
From ICT to computing

Eleanor Overland

Introduction

Computing is a relatively new subject area compared to the well-established curriculum subjects such as mathematics. There is not yet the same body of research or maturity in thinking of the subject as a bounded body of knowledge. To problematise the study of computing further, the subject is ever-evolving. With the advent of new technologies, changes in software, programming languages, protocols, ethical questions, etc., a computing curriculum is always a work in progress.

When starting to think about the computing curriculum, I encourage you to engage in some personal reflection. Your own personal experiences and developing interest in computing education form part of the story of computing development. Your personal educational experience illustrates a part of that journey. Did you experience computing at school? Was it even called computing? Did you study it at higher levels? Have you worked in the computing industry and if so, what led you to choose that path? Who have you met along the way? What equipment or software have you used? How did you build your own computing knowledge? Did you find your own experiences of computing education to be balanced and inclusive of all? As you read this chapter, you will probably revisit these questions and many more as we consider the changes to the computing curriculum and current considerations to ensure it is inclusive for all learners.

Personally, I did not study computing (or any other related subject at school). It was not on offer to me. (Despite the views of my own children, I am not ancient!) I was not at school in a pre-computer age; it was just that computing was not a curriculum priority. There were Information Technology (IT) suites, and we could occasionally use them if we had the suitably enthusiastic maths teacher who was confident to take his classes in there. I think he ran a computing club that some of the boys went to, and I am aware of some girls who learned word processing as one of their options, but that was my only real experience of computing at secondary school. It was only at university I discovered my enjoyment of working in computing. I was fortunate enough to have (unintentionally) chosen a course with technology-rich modules, which sparked my interest and ultimately led to my choice of career. I sometimes wonder how many other children from my school would have enjoyed working with computers and if they have managed to find their own way into it since.

Since my own schooling, my experiences as a teacher also say a lot about the rate of change and levels of inclusion in the subject. I have taught a huge range of information communication technology (ICT), IT, computing and digital qualifications with more acronyms and initials than I care to remember. As a very dull party trick, I can work out someone's age based on the ICT qualifications they have (if you have a certificate in digital applications (CiDA), I am guessing you were born in the late 1990s or early 2000s'; iMedia are the real youngsters, whilst those with computer studies are the more mature readers!). What is far more interesting than "what" I have taught is "who" I have taught. I have taught general certificate of education (GCSE) ICT classes with a great mix of children reflecting the wider population of the school. I have also taught vocational qualifications full of lower attaining learners where their Head of Year has advised them it is the most suitable option for them. More recently, I have taught GCSE computer science classes where I have been the only female and non-white person in the room.

Since I have moved into teacher education, I am pleased to say I have witnessed further positive change with many schools working hard to improve the inclusivity of the subject. Some schools have struck a balance between genders, socio-economic representation, ability range and race within their classes. However, many schools are still facing challenges with regard to inclusive computing education. As computing teachers, you will have an opportunity, a duty even, to understand why some of these differences exist and how your own curriculum design and teaching can influence inclusivity in the subject.

Subject and curriculum development

Personal computers started to appear in mainstream consciousness in the 1970s but did not start to appear in schools until the early 1980s (Somekh, 2007). As a school "subject" compared to such realms as mathematics and geography, this makes it a relative newcomer to the curriculum subjects delivered in schools. Many early policies and shifts towards using technology in the classroom were funding and technology-driven rather than curriculum and qualification driven, particularly with the introduction of the BBC Micros and funds given to the school for the purchase of hardware and software annually by the Department of Trade and Industry (Somekh, 2007). Initially, there were very few teachers with any formal training in using the microcomputer in school. Many teachers were self-taught and school leaders relied on the enthusiasm of individual teachers to develop their usage in school. Due to the required mathematical understanding of making use of the Micros, it was often a mathematics or physics teacher within schools who felt an infinity with computing development. Computer studies at CSE and "O" level (qualifications at school leaver age, a precursor to current GCSEs) were required to make use of the first PCs through formal study of programming and computational thinking. Many of the theoretical and programming-based questions would not look amiss in the newly developed Computer Science GCSEs designed to assess the latest computing curriculum (Simmons & Hawkins, 2015).

The Technical and Vocational Education Initiative (TVEI) qualifications in the early 1980s were the first group of qualifications to recognise IT skills. These were designed to be a vocational route for those pupils less likely to gain more formal, academic qualifications but to ensure this group of learners were prepared for working life with up-to-date skills (Williams, 1993). These qualifications developed skills deemed necessary within the workplace at the time such as word processing and data entry. During this period, many schools also offered the opportunity to develop touch-typing skills through qualifications such as the Royal Society Arts (RSA) typing award (Hillier, 2012). With these qualifications being closely aligned to work-based learning, they required the skills of a range of teachers including those specialising in business studies or technology. Vocational qualifications such as these were gained by some of the most disadvantaged learners in school who may not have otherwise secured formal academic qualifications at this stage of their education.

In the early times of use of computers within schools, the notion of a specialist teacher was not someone with relevant training or qualifications in the area, but someone willing to give it a go and "learn on the job". Much concern has been raised in more recent reviews and studies regarding the lack of non-specialist teachers within the discipline (Royal Society, 2012), but it is seldom acknowledged that the whole subject was developed by "non-specialist", enthusiastic frontrunners who now may well be the most experienced educators in the field.

In the late 1980s, the subject acquired an umbrella term IT. The growing prevalence of the internet in the 1990s added the "Communication" aspect to ICT. Vocational qualifications continued to develop, with General National Vocational Qualifications (GNVQs) in ICT superseding the TVEIs. Qualifications developed for use in the workplace were also adopted by some schools such as Business and technology education council (BTEC), computer literacy and information technology (CLAIT) and the European Computer Driving Licence (ECDL). Many of these level 2 qualifications carried the same value as GCSE passes at Grade C or above so became popular with school leaders as school performance measures became critical. A full-level 2 GNVQ in ICT carried the same equivalence of 4 GCSEs in school performance measures despite being based on competence measures rather than examinations and being delivered in a much reduced amount of curriculum time. New teachers were starting to appear who had computing experience from industry or their own university studies, and the New Labour Government launched the National Grid for Learning with over £700 million pledged to schools to purchase hardware, software and internet connections. Along with this, a £230 million "New Opportunities Fund" (NOF) was announced to pay for training all teachers and many school support staff in ICT (Somekh, 2007). In 1999, ICT was given status as a National Curriculum subject in its own right with programmes of study for pupils to follow from infant education through to age 16. Compare the £230 million NOF money to the £84 million announced in January 2018 and, even without allowing for inflation, it is clear that the late 1990s was a boom time for investment and growth in ICT education in schools.

Computer studies was scrapped with the removal of CSEs and "O" levels in 1988 with very little of the content making it to the new GCSE ICT qualifications. Initially, many schools preferred to make use of the fruitful vocational qualifications on offer but, with the addition of English and Maths as a measure of the 5 A-C success in schools (Parameshwaran & Thomson, 2015), in many cases, the subject was given a reduced curriculum time and also carried less status with school leaders, parents and pupils. Initially, there was considerable uptake for GCSE ICT as ICT became a core subject at KS3 and so a natural progression for pupils was to the KS4 qualification. In 2007, statutory assessment tests (SATs) in ICT were piloted for the first time for 13 year olds, although these were scrapped shortly afterwards due to the complexities of running such a large scale, online practical examination.

In the mid-to-late 2000s, a decline in GCSE ICT started. GCSE entries in 2008 were approx. 68,000, dropping to 32,000 by 2011. This is due to some pupils following vocational courses but also a perception by children or parents that their regular use of computers outside of the subject reduced the need to study it as a specific subject (Royal Society, 2012). Despite the decline, the gender balance remained stable with approx. 45% of entries from female learners (Ofqual, 2021). The initial rhetoric that ICT qualifications would be essential for any school leaver seemed to carry less clout, and as computers were being used widely by pupils in other aspects of their lives, parents seemed less concerned with a formal education in the subject. Over this time, the required content for the ICT qualifications changed very little, apart from being updated for later versions of software, and curriculum time for the subject was being ever more squeezed to make way for other subject areas (Ofsted, 2013). At a similar time, a grassroots organisation, with the support of the British Computing Society (BCS), was formed, called "Computing at School" (CAS). They started with an initially small membership to introduce computing focussed rather than ICT focussed education in schools, and a pilot GCSE in computing was first developed in 2011. It was the slow decline of ICT and a concern for the skills and understanding of the future workforce that also prompted to call for a review of ICT education, carried out by the Royal Society, led by Professor Stephen Furber.

The "Shutdown or Restart" report commissioned by the Royal Society (2012) was pivotal in presenting a view of delivery of the then ICT curriculum, now computing education within UK schools, and this has since been used to guide and influence curriculum policy.

The Shutdown or Restart report made a number of recommendations for the renewal of the ICT curriculum. The first was a total overhaul of the National

Curriculum, rebranding ICT to be computing and incorporating IT computer science and digital literacy. The report also concluded that many teachers lacked the knowledge and expertise required for the new rigorous curriculum and recommended a range of continued professional development (CPD) to be funded for teachers. The report recommended these fundamental changes to give the subject academic rigour and improve the number of learners opting to study it at higher levels (Royal Society, 2012).

In 2012, the then education secretary, Michael Gove, made a crucial speech at the British Educational Technology Exhibition (BETT) following the Royal Society report. During the speech, he outlined a number of criticisms of the then still compulsory National Curriculum in ICT, stating concerns about its lack of stretch, opportunity for creativity and it being generally dull and off-putting. During the speech, the ICT National Curriculum was disapplied with immediate effect, announcing a freedom for all teachers in the field to cover innovative, specialist and challenging topics (Gove, 2012). He made it clear that the government must not wade in and prescribe to schools exactly what they should be doing or how they should be doing it. Less than a year later, the National Curriculum programmes of study in computing were published (DfE, 2013).

GCSEs in computing were developed and rolled out by all major examination boards across England and Wales from 2012 onwards. Under the GCSE reforms, in September 2016, the qualifications were changed to "computer science" to provide additional academic rigour and to meet the new GCSE 9-1 criteria (Ofqual, 2018). CAS developed the "Network of Excellence", supported by funding from the DfE. The network has created local coverage of "hubs" to support computing teachers at a local level, an online platform for forums and sharing of resources and a range of CPD provided by "Master Teachers", specialists who are funded to come out of schools for short periods to run CPD for developing teachers. GCSE ICT examinations were sat for the last time in the summer of 2017, and data show a small but steady increase in the numbers of pupils being entered for GCSE computer science.

The current position in computing

The changes in examination entries and schooling during the pandemic has skewed data around examination entries and performance. The most reliable recent data are pre-pandemic with the Roehampton Annual Computing Education Report (2018) drawing findings from performance tables for examinations and the school workforce report (Kemp & Berry, 2019). In 2018, approximately 12.4% of all pupils sat GCSE computer science. Of these approx. 20% of candidates were female, which is the lowest proportion for any science technology, engineering and maths (STEM) subject (WISE Campaign, 2018). Whilst the report identified the number of schools offering access to GCSE computer science had increased, access was greater for pupils in more affluent areas and only 4% of special schools offered the qualification. The subject research report (Ofsted, 2022) identifies gender divides appearing earlier than key stage 4, with girls in year 9 already more likely than boys to describe the subject as boring and to have less confidence in their own abilities in the subject. Despite the lack of confidence, girls do outperform boys at GCSE level, although all children entered for the qualification underperform in GCSE computer science compared to their other subjects.

Concerns in the success of computing as a subject are not only reflected in pupil data but also in broader measures including the availability of specialist staff and a gradual reduction in curriculum time. In fact, the implementation of the new computing curriculum is stated as being "patchy and fragile" in the "After the Reboot" report (Royal Society, 2017). The report was a follow-up by the Royal Society to review the implementation of the new curriculum. Many of the findings and recommendations focussed on improving levels of access and inclusion within the subject. This is mirrored in the Ofsted research report, and it is clear that there is room for development to ensure the computing curriculum is accessible for all.

Think about computing within your setting. What is the current position? Does it reflect the national picture summarised here or are there differences? Does the current position reflect where you believe it should be? Some key questions to think about here are whether all children know and remember the computing knowledge you feel is important? Do all children have opportunities to study computing to a level that prepares them for further education, training or employment? Of those securing high-quality qualifications in computing, are some groups of children underrepresented? Thinking about your own answers to these questions can be a useful starting point for thinking about curriculum design and inclusive practice.

Designing an inclusive digital curriculum

New and early career teachers may think that they do not need to be concerned with curriculum design. They tend to be more focussed on short-term planning and individual lessons. This is an underestimation, firstly not only of the importance of the bigger picture but also of the potential for influence and debate those new teachers have as part of the profession. Schools are increasingly placing the curriculum at the heart of what they do, focussing on the curriculum's intent, implementation, impact and level of ambition for their children (Ashbee, 2021). Taking time to consider some of the most fundamental questions in education and being up-to-date with the latest research is in the domain of new teachers. It is a common interview question for new teachers, "how would you design the curriculum in your subject?" So it is an important question to consider, but there is no single or simple answer.

Under the Ofsted framework, schools are expected to outline the intent and implementation of their curriculum for each subject area. They are also expected to be ambitious for their learners. Ashbee outlines four moral purposes for having an ambitious curriculum, each of which highlights the importance of an inclusive curriculum accessible to all learners. I will take each in turn and consider it from a computing perspective.

The cognitive purpose of the curriculum raises the importance of knowledge. In computing, this involves both declarative and procedural knowledge, not only remembering facts and information but being able to carry out practical activities, processes and skills. The moral argument for ensuring an inclusive curriculum is that all children should have the opportunity to secure this knowledge. It requires careful sequencing to ensure knowledge is "built" in the appropriate order to ensure that children are able to make connections and commit knowledge to their long-term memory. In computing, it requires a skilful sequencing of the curriculum to ensure learners are able to build on prior knowledge in a meaningful way, for example, too much focus on programming language syntax prior to an understanding of the structural concepts of the program being built may lead to superficial knowledge of programming, which learners are then unable to retain or apply.

The socioeconomic moral purpose is strong in the computing curriculum. The developments of the subject described earlier in the chapter are driven by the developments of the technology at the time and ultimately the job market. Eric Schmidt, the then CEO of Google, was influential in changing the landscape from ICT to computing citing the need for younger generations to be prepared for new technology-based industries (Gove, 2012). Computing knowledge is also applicable to many careers, not those solely based in digital industries. For example, many of the creative industries rely on digital knowledge and skills in their practice. For any learners who do not have equitable access to the computing curriculum, they may find themselves at considerable disadvantage in employment, regardless of their career choice.

The lack of inclusion in access to the curriculum can also have socioeconomic implications for businesses. There is already a widely acknowledged shortage of digital skills in areas of the UK. The annual digital skills audit 2021 identified that 20% of companies involved in the survey were unable to fill their digital or technical roles, mainly due to a lack of quality candidates. The same audit also identified that only 26% of the workforce in technical roles were female. Of all the employees, 87% were white and 11% identified as having a disability or long-term condition (UK Tech Cluster Group, 2022). Greater inclusion in computing education would contribute to a larger and more skilled digital workforce. An under-representation of groups in the workforce also potentially prevents products from meeting the needs of different consumers in the market. An example that took the media by storm was the case of the Apple Health app. It was only in 2015, after feedback from thousands of women, the app introduced features to allow women to track their menstrual cycle. The company admitted it had not thought of this at the initial development stage. At the time, only 20% of developers at Apple were women (Griffin, 2015).

The third moral purpose is critical in computing, and this focuses on democracy. This explores the situation where children need to have a certain level of knowledge in order to participate in a healthy democracy. This ranges from learners being able to recognise fake news to having a level of critically around policy and behaviours such as ownership of data. Chapter 4 explores this in much more detail exploring the bias in artificial intelligence. Without powerful knowledge, it is difficult to make fully informed decisions about the use of technology. Current learners are going to be engaged in ethical debates as the development of technology-driven applications and automation becomes more widespread. Without an inclusive curriculum, we are potentially excluding groups from future democracy.

The final moral purpose for an ambitious curriculum is to facilitate intellectual development. This allows children to make connections between their learning, make meaning in the world and generate new ideas. Ashbee (2021, pp. 12) argues that this should be the entitlement of every child, from every household and every background. With one of the advantages of intellectual development being the ability to pass on knowledge to future generations, there is a risk that any underrepresentation could become cyclical with certain groups "excluded" for generations to come.

When thinking about curriculum design, it may be useful to reflect on your own beliefs and vision for the computing curriculum. What is it you want learners to gain from studying computing? Have you considered any moral or economic drivers when thinking about curriculum design? The key question when thinking about inclusion is whether it is accessible and purposeful for all learners. That is at the very heart of inclusion.

Building of knowledge

The purpose of the curriculum (the intent) provides a starting point from which to determine required knowledge and appropriate sequencing. The National Curriculum provides a clear framework for teaching curriculum subjects, but within that framework, there remains a freedom for teachers to develop and so realise the specific intent of their curriculum. One approach to doing this is to consider knowledge in computing and determine what it is the learners most need to know and remember. Muller and Young (2019) describe powerful knowledge as that which gives greatest potential for explaining the world and new ways of thinking about it.

The framing of powerful knowledge is a useful way to consider inclusion in the curriculum, as many of the parameters for determining whether knowledge is "powerful" draw on questions around social justice and inclusion (Didau, 2018). Children need powerful knowledge so that they do not remain dependent upon those who have it. Shared powerful knowledge allows children to grow into useful citizens and it "opens doors" so should be available to all children. A particular feature of powerful knowledge is that it should be determined within the subjects and should be cognitively superior to that to that needed for daily life (Didau,

2018, pp. 205). In computing, this emphasises the need to take learning beyond the everyday use of computers to having an understanding of fundamental computing concepts. This raises a particular question for children not opting to study computing-related qualifications. That is, is there somewhere else within the curriculum they are securing powerful knowledge in computing?

Muller and Young (2019) emphasise that powerful knowledge will take students beyond their own experiences but should also relate to their own context. One way to take learners beyond their own experiences is through development of cultural capital. Ofsted (2019) cite cultural capital as an essential aspect of any curriculum design. If children are exposed to more opportunities, they will improve their understanding and success in school and beyond. To improve performance of different groups, particularly disadvantaged children, schools are increasingly looking to improve access to enriching activities within their curriculum. Many descriptors of cultural capital tend to be quite traditional, involving trips to theatres, music lessons, outdoor activities and the like. Whilst there are many opportunities in computing such as a trip to Bletchley Park or an after-school coding club, there is an opportunity to think of cultural capital in computing in a more nuanced way.

Cultural capital has its origins in in the 1970s with the sociologist, Bourdieu. Bourdieu outlined three categories of cultural capital, objective (cultural goods, books and works of art), embodied (language, mannerisms and preferences) and institutionalised (qualifications and education credentials) (Bourdieu, 2010; Reid, 2020). Now reimagine these as computing-based concepts linked mainly, but not exclusively, to the digital literacy aspects of the national curriculum. In terms of objective cultural capital, consider pupils access to and understanding of online resources. Are children able to be discerning when sourcing reliable online content, especially news and current affairs? Are they able to pay for high-quality apps (such as revision organisers) or purchase quality online content? Considering embodied cultural capital, do the children know how to be professional online? Have they created high-quality blogs or social media threads? Do they know how to search for jobs, do online banking and be prepared for adult life online? In terms of institutionalised cultural capital, do all the qualifications on offer as part of your curriculum carry the same level of cultural capital? For children who do not opt for qualifications at GCSE level, are they still able to access computing courses, perhaps online, or gain recognition (such as digital badges) for developing computing skills? With all three aspects of cultural capital, families can have a real influence on the behaviours and access for children. To be fully inclusive, it is important to think about all learners and how they may gain such experiences and access through the school curriculum.

One of the reasons cultural capital is foregrounded by Ofsted is the recognition of the importance of families and the impact of learning outside school has on pupil progress. This was never more apparent than during the national lockdowns in 2019 and 2020. As children were homeschooled, there was a rush to get devices and Wi-Fi access to families (Greenhow et al., 2021). What was less easy to address was the level of support for parents and carers (Manca and Delfino, 2021). Community organisations quickly realised that, although devices were issued, some children were still unable to access online lessons. They quickly produced guides in different languages to support families where English was not the home language. Schools worked hard to provide videos and help sheets to try and make it as simple as possible for parents and carers to support children. Technology provides a wealth of inclusive tools such as screen readers, language translators and additional educational resources, but knowing these tools exist and how to operate them can be far more problematic than simply having access (The Greater Manchester Technology Fund, 2022).

In many schools, the pandemic has increased the use of online teaching resources, especially for homework, and this has continued as schools have returned to more normal operations. Despite this, the issues around access and support at home are still there. When considering inclusion in your curriculum design, is this an issue? Do the children who struggle most in lessons also struggle to access technology and/or the internet at home? Are they not the children who most need to access additional learning materials? One suggestion I often hear is that children can access it at lunchtime or after school. Again, this asks the question of whether or not this is an effective inclusive practice.

Conclusion

In this chapter, I have tried to outline the many potential areas where the current provision in computing education is failing to be inclusive. A main issue is that children have just not managed to identify with the subject, they cannot position themselves within it or do not see the relevance of the subject in their future lives. You have the opportunity to address these issues within your own teaching through ensuring that computing education is inclusive.

A second particular issue is that learners are finding the subject too difficult. Their knowledge has not been sequenced in a way to build understanding, confidence and growth. Again, as a computing teacher, this is something you can develop within your own practice.

The chapters that follow in this book will provide some real insights into inclusive practice in computing education. Some chapters may influence your choice of taught topics, sources of teaching materials or opportunities to build cultural capital. Other chapters may influence your classroom practice, pedagogical approaches and the sequencing of computing knowledge. This book, however, is not exhaustive. Research in computing education is ongoing. As you continue to engage with the latest publications and be part of the community of computing educators, your practice will continue to evolve. You will also find things out yourself in your own classroom. Reflecting on your own classroom practice, analysing pupil progress and student voice activity will help inform your own approach to inclusive computing education.

References

Ashbee, R. (2021) *Curriculum: Theory, Culture and the Subject Specialisms*. Routledge. Abingdon.

Bourdieu, P. (2010) *Distinction: A Social Critique of the Judgement of Taste*. Taylor & Francis Group. London.

DfE (2013) National curriculum in England, computing programmes of study. https://www.gov.uk/government/publications/national-curriculum-in-england-computing-programmes-of-study.

Didau, D. (2018) *Making Kids Cleverer: A Manifesto for Closing the Advantage Gap*. Crown House Publishing. London.

Gove, M. (2012) Michael Gove gives a speech at the BETT Show 2012 on ICT in the National Curriculum. http://webarchive.nationalarchives.gov.uk/20131212112355/https://www.gov.uk/government/speeches/michael-gove-speech-at-the-bett-show-2012.

Greenhow, C., Lewin C. and Staudt Willet K B. (2021) The educational response to Covid-19 across two countries: A critical examination of initial digital pedagogy adoption. *Technology, Pedagogy and Education* 30(1):7–25.

Griffin, A. (2015) Apple's health data app gets period tracking. The Independent. https://www.independent.co.uk/tech/apple-s-health-data-app-gets-period-tracking-10307363.html.

Hillier, Y. (2012) *Reflective Teaching in Further and Adult Education*. 3rd edn. Continuum. London.

Kemp, P.E.J. and Berry, M.G. (2019) *The Roehampton Annual Computing Education Report: Pre-release Snapshot from 2018*. University of Roehampton. London. https://www.bcs.org/media/2520/tracer-2018.pdf.

Manca, S. and Delfino, M. (2021) Adapting educational practices in emergency remote education: Continuity and change from a student perspective. *British Journal Education Technology* 52:1394–1413.

Muller, J. and Young, M. (2019) Knowledge, power and powerful knowledge re-visited. *The Curriculum Journal* 30(2):196–214.

Ofqual (2018) Future assessment arrangements for GCSE (9-1) computer science. https://www.gov.uk/government/consultations/future-assessment-arrangements-for-gcse-computer-science.

Ofqual (2021) GCSE outcomes in England. https://analytics.ofqual.gov.uk/apps/GCSE/Outcomes/.

Ofsted (2013) ICT in schools: 2008 to 2011. https://www.gov.uk/government/publications/ict-in-schools-2008-to-2011.

Ofsted (2019) Education inspection framework. https://www.gov.uk/government/publications/education-inspection-framework.

Ofsted (2022) Research review series: Computing. https://www.gov.uk/government/publications/research-review-series-computing.

Parameshwaran, M. and Thomson, D.J. (2015) The impact of accountability reforms on the Key Stage 4 curriculum: How have changes to school and college Performance Tables affected pupil access to qualifications and subjects in secondary schools in England? *London Review of Education*.

Reid, A. (2020) Cultural capital, critical theory and curriculum, in Sealy and Bennet (eds.) *The Researched Guide to the Curriculum: An Evidence Informed Guide for Teachers*. John Catt Education. Woodbridge.

Royal Society (2012) *Shut Down or Restart? The Way Forward for Computing in UK Schools*. The Royal Society. London, UK. https://royalsociety.org/topics-policy/projects/computing-in-schools/report/.

Royal Society (2017) *After the Reboot: Computing Education in Schools.* The Royal Society. London, UK. https://royalsociety.org/-/media/policy/projects/computing-education/computing-education-report.pdf.

Simmons C. and Hawkins C. (2015) *Teaching Computing.* SAGE. London.

Somekh B. (2007) *Pedagogy and Learning with ICT: Researching the Art of Innovation.* Routledge. London.

The Greater Manchester Technology Fund (2022) The Greater Manchester Technology Fund – What we do. https://www.greatermanchester-ca.gov.uk/what-we-do/digital/get-online-greater-manchester/the-greater-manchester-technology-fund/.

UK Tech Cluster Group (TCG) (2022) Digital Skills Audit 2021 https://www.manchesterdigital.com/post/manchester-digital/skills-audit-2021-national?submitted.

Williams R. (1993) The fate of the technical and vocational education initiative in a pilot school: A longitudinal case study. *British Educational Research Journal* 19(4):421.

WISE Campaign (2018) https://www.wisecampaign.org.uk/statistics/analysis-of-gcse-stem-entries-and-results-2/.

The development of artificial intelligence in computing education

Thinking betwixt and between – reinvigorating Papert's im/possibilities of computing

Amanda Banks Gatenby

Introduction

Fifty years ago in 1972, Seymour Papert and Marvin Minsky wrote a progress report for the artificial intelligence (AI) lab at the Massachusetts Institute of Technology, on technical progress and their "point of view about certain problems in the Theory of Intelligence" (Minsky and Papert, 1972). Through their attempts to represent computationally the processes of intelligence, Papert and Minsky explored what intelligence was not as well as what aspects of it might be possible and not possible to model with a computer. This pre-dated Papert's increasing focus on embedding computational perspectives in schools and other formal education settings.

Their report begins with the recognition that, "Our evolution of theories of intelligence has become closely bound to the study of development of intelligence in children". Current research in the area of developmental and evolutionary robotics similarly models the processes of early learning with embodied AI, acknowledging that the *process* of learning itself changes over time (for example, Cangelosi and Schlesinger, 2018). Papert's work was equally revolutionary in recognising that not only do the learning processes change, but the 'stuff' being learned also changes through the process of learning it (Noss and Hoyle, 2013). For Papert, a computer offers the medium and the literacy that can help us deconstruct, entangle and remake our understanding of what it means to learn. Through his well-known educational works including *Mindstorms* and *The Children's Machine*, Papert's advocacy of computers in classrooms was to explore how learning to represent ideas computationally could teach us more about how we learn and think.

In the following ten years after the AI progress report, Papert's attempts to integrate this perspective into schools through his work with the Logo programming language and associated tool, the Logo Turtle, did not have the intended outcomes. Papert became increasingly critical of schools and wrote an article in the Times Educational Supplement in 1982 in which he stated, "I think schools generally do an effective and terribly damaging job of teaching children to be infantile, dependant, intellectually dishonest, passive and disrespectful to their own developmental capacities" (Papert, 1982). Agalianos, Noss and Whitty (2001) describe the issues that led to unforeseen impacts of the Logo project and explain that the essential problem was:

> ... the gap between the initial expectations and the reality of its implementation demonstrates that the technology needs to be surrounded by social and political relationships that will allow it to do transformational work; that the introduction of technology alone cannot possibly bring about radical change; that the medium alone cannot carry the entire message.
> (Agalianos, Noss and Whitty, 2001)

However by the 2000s, Papert was once again writing optimistically about the changes in schools being driven by the development of new technologies, through his Piagetian lens. In this view, schools had the potential to change in the same way that children change through learning (Papert, 2005). In particular, there would be wider forces than teachers' intentions in classrooms that could enable "mega-change" to emerge, through more general developments in technology (Papert, 2000, p. 728). Any current discussions of a "mega-change" in education usually involve mention of AI, the field in which Papert's interests in learning really began. As AI technologies become more common in everyday discourse and as debates increase around the use of AI both in and for educational purposes, there is recognition about the importance of enabling young people to take a critical stance on these technologies (see for example, Luckin, 2018; The Institute for Ethical AI in Education, 2021).

Including criticality in computing education is not new with the *Shut Down or Restart* report of 2012, claiming that learning about computing was a question of empowerment and participation, ensuring, "Citizens able to think in computational terms are able to understand and rationally debate issues involving computation" (p. 29). Zook and Graham (2007) claim the "control of code is power" and Mitchell asks us to consider:

> Who shall write the software that increasingly structures our daily lives? What shall that software allow and proscribe? Who shall be privileged by it and who marginalized? How shall the writers of the rules be answerable?
> (Mitchell, 1996)

Those who are or who will be the creators of our technologies, which increasingly determine the form of our practices, have considerable power, and we would hope that they have the critical thinking abilities to underpin the design and building of those technologies, to have awareness of these implications Mitchell raises such as who may be marginalised by a particular tool, for example. But the notion of criticality is particularly complex in computing education. Once students reach higher education (HE), those in computing and engineering disciplines are "used to sitting and listening... learning the correct way to do things" (Bullen, 1998, p. 30). But if students' underlying perspective is based on a deterministic "correct way", this is problematic. For example, Chan, Ho and Ku's (2011) research with students in HE settings suggests that students who view knowledge as certain have a reduced capacity for critical thinking.

Much of the focus on the turbulent changes in curriculum in the last ten years following the publication of the *Shut Down or Restart* report has been on the term of *computational thinking*, and in Banks Gatenby (2017a, 2017b), I question whether existing frameworks of "computational thinking" and critical ways of thinking are a pedagogical antithesis or whether it is possible for computing teachers to incorporate both into their pedagogical landscapes and if so, how they might do this. Frameworks such as Brennan and Resnick's (2012) offer a more promising approach with a focus on concepts, practices and perspectives. But the way in which tools such as Scratch are implemented in teaching can often be a rote, "death by Scratch" (Sentance and Csizmadia, 2017) approach, reinforcing the perspective that there is one right way to do things and missing opportunities for creativity and criticality in problem solving.

As educators consider ways to integrate AI into computing curriculum and pedagogy, how might Papert's powerful ideas, which were largely influenced by work in AI, be revisited? A recently developed tool to explore this with is Dale Lane's Machine Learning for Kids (MLfK) at: https://machinelearningforkids.co.uk/. Built on IBM's Watson and integrating with Scratch, the tool has a range of worksheets detailing projects for students. There are pre-trained models that have extension blocks which can be added to Scratch but students can also create and train their own models in ways that demonstrate fundamental ideas about machine learning. The rest of the chapter will draw on this and other tools to ask how we might reinvigorate the foundational themes of Papert's work to help computing educators more powerfully put to use the subject of powerful ideas (Papert, 1980).

Building on Piaget and Vygotsky – computation for unmaking concepts

Papert is most closely associated with the learning theory of constructionism which is considered to extend Piaget's work to include artefacts. This involves the construction not just of physical artefacts but of new ideas, with the emphasis being on the intention of sharing those artefacts with others. For Papert, the medium of computing offered the possibility of turning education on its head from a transference of existing knowledge from adult to child, to all learners creating new representations, forms of ideas and new knowledge together through this machine that sits "betwixt and between the world of formal systems and physical things" (Turkle and Papert, 1990). In this possibility lay a fundamental shift in inclusivity in teaching and learning where all perspectives could be valued. From his very early works with children, Papert believed this was something all children could learn to do, not just those with an "aptitude", if taught in particular ways (see for example, Papert, 1972).

Papert moved from completing his work on two PhDs in maths to modelling knowledge processes with AI techniques and advocating for a new understanding of the benefits "computational doing" could bring to schools, learners and our understanding of the learning process. Logo was a tool to support this project, but the unintended consequence of the take up of Logo in schools led to the situating of computing as a subject for its own sake, in many ways the very opposite to Papert's intentions. Papert saw a computer as offering a personalised learning experience, by which he meant a means for pupils to search for solutions to immediate problems and in the process, discovering what works for them.

This perspective of "personalisation" is not the "chocolate coated broccoli" (Noss and Hoyle, 2013) approach that dresses up what is being taught in a superficial topic with the appearance of being more personalised; for example, spreadsheets might be taught through a task based on football league tables. For Papert, "meaningful" or "personalised" learning doesn't mean a vague connection through the topic to a hobby, which is of interest to some of the class. Neither does it mean automating the selection of a task based on how accurately a prior task has been completed, as is the basis for many current data-driven approaches that claim to offer personalised learning (see chapter 5). This latter approach depends heavily on the efficacy of the design of the previous tasks and in that sense is instead inherently impersonal. For Papert, a meaningful task is immediately (both spatially and temporally) useful, having a purpose through a project or problem that needs solving here and now. In this way, the necessary tools or approaches are discovered because they are needed to solve the problem at hand.

Many of the worksheets in MLfK have the potential to relate to meaningful and immediate needs of young learners and additionally provide a basis for young people to engage in wider discussions around current technological developments. For example, there is a chatbot project in which students can choose their own topic to be the focus of communication, a virtual school "librarian recommender", a "smart classroom" project and a Pokemon predictor. There is also a project to program the computer to recognise the difference between a kind comment and an insult (see Figure 2.1).

Crucially, in the process of solving a problem, the student can learn about *how* they think and learn. Papert suggests that some of the most interesting projects

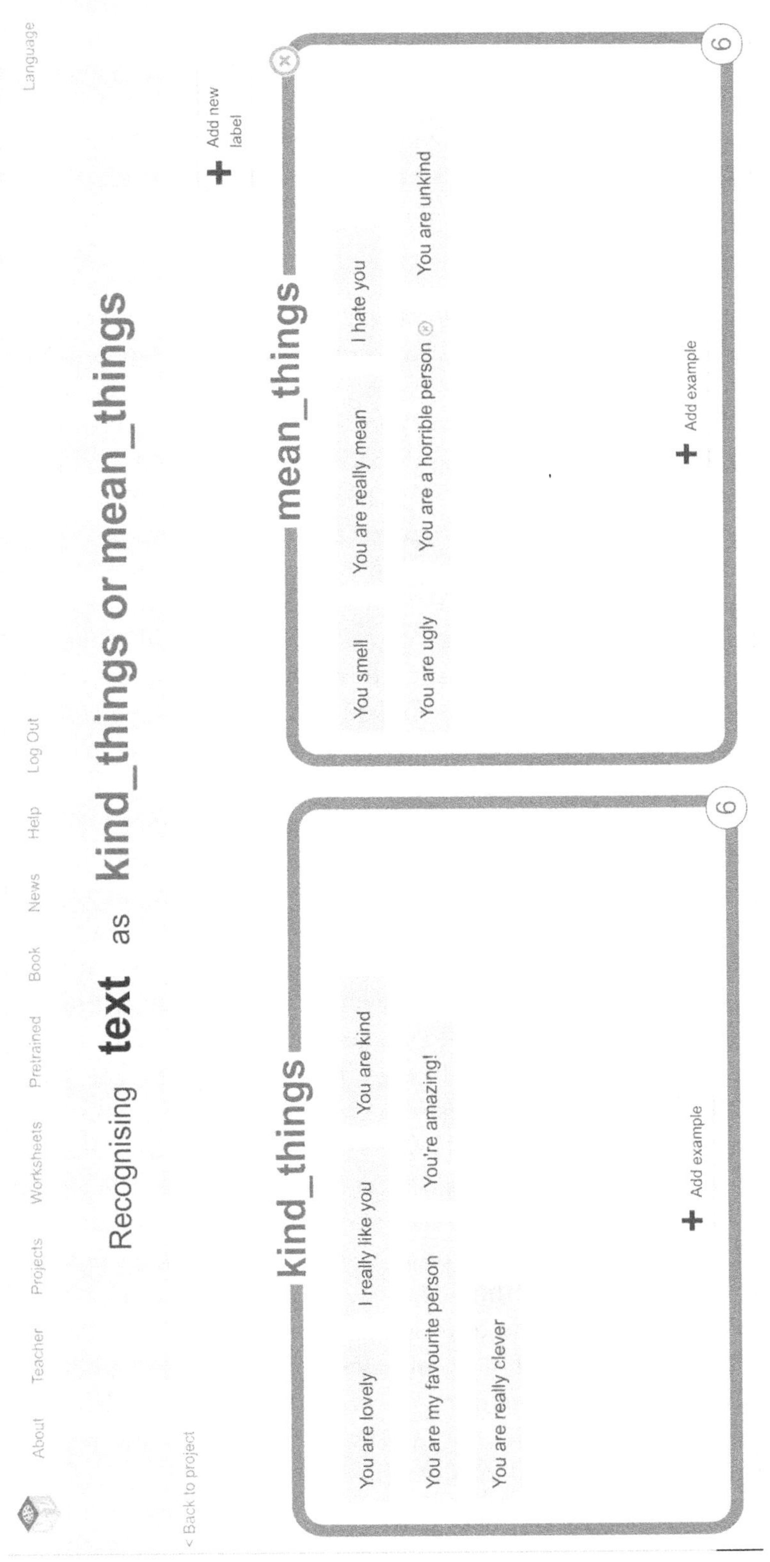

The development of artificial intelligence in computing education 23

About Teacher Projects Worksheets Pretrained Book News Help Log Out   Language

< Back to project

Recognising **text** as **kind_things or mean_things**

+ Add new label

─── kind_things ───

You are lovely   I really like you   You are kind

You are my favourite person   You're amazing!

You are really clever

+ Add example

6

─── mean_things ─── ⓧ

You smell   You are really mean   I hate you

You are ugly   You are a horrible person ⓧ   You are unkind

+ Add example

6

Figure 2.1 Screenshot of model training with machine learning for kids.

he explored were in solving teaching-related problems. He describes examples of children who were alienated from learning maths who were thoroughly engaged in writing programs to teach maths to their peers and younger learners:

> It is said that the best way to learn something is to teach it. Perhaps writing a teaching program is better still in its insistence on forcing one to consider all possible misunderstandings and mistakes.
>
> (Papert, 1972)

To understand more about how computation can support this, it is helpful to explore the influence of Piaget on Papert and how his work can be seen as more closely aligned with Vygotsky's thinking. In Papert's later work with Sherry Turkle, he critiques Piaget in terms of his situating the acquisition of "formal thinking" as a developmental stage. Rather they say it is a style, or a tool, (Turkle and Papert, 1990, p. 10) which is more in line with Vygotsky (for example, 1986). Vygotsky gives primacy to social interaction, just as Papert gives primacy to the environment and objects within it.

In *Thought and Language*, Vygotsky expands on the notion of "concept" and how a potential concept develops into a "true concept". For Papert, the computer bridges abstract or formal systems and concrete, physical things (Turkle and Papert, 1990, p. 2). Exploring computation through a Papertian lens offers both an example of how computation helps us understand a notion such as "concept" by unmaking it; and how we can develop our understanding of the learning process by attempting to represent a simple concept.

Taking the example of a simple concept such as *square,* we can imagine the complexity of how this concept develops over time, starting with a baby sitting on a carpet playing with square blocks. They hear the sound of a parent or carer saying "square". Every time they play with the blocks, they hear that sound and hear the same sound when in the presence of other *squares*. They learn that when in the presence of circles, the sound "square" is not used, thus over time that sound "square" gets associated with the child's experiences of the shape. Once able to physically use a pencil, the child might learn to draw a square, and now they understand that four lines in a certain pattern on a page can represent *square*. They begin to learn how to read, how the character "S" represents the sound "sss", and over time, they learn that characters "s", "q", "u", "a", "r" and "e" in sequence, are another way to represent and communicate about *squares* – the letters representing the word and its sound "square", which is associated with experiences of *square*. As they enter formal education, they might learn more formal (in Vygotsky's theory this would be "scientific" as opposed to "spontaneous") understandings of square – "it is a shape with four sides of equal length". This is added to and entangled with that child's understanding of *square* without removing in any sense the already unimaginably complex set of unique experiences and representations by which the child currently understands the meaning of a "simple" concept such as *square* (we're ignoring the multitude of relationships between this and other concepts). These understandings are unique to that child, due to their multitudinous personal experiences of *squares* encountered, but they are fundamentally social, in that the experiences and representations of them are given by and shared with the environment, including other people.

Papert argued that computational literacy is a new way for learners to express concepts and contributing to how they are understood and developed further. To program such a representation or model, we have to understand the process by which we can create such an entity. If students are asked to program a square, using a tool like Scratch, they have to think about how to *move* in a square, or the *process* of making a square, rather than simply the features of the object – they code *how to get to a square*. This makes explicit the connections between the abstract, idea of *square* and the physical experience of *square*. They would use directions something like:

Move FWD 10

Rotate 90 degrees

Repeat 4 times

This is the process of computationally drawing a square and gives another representation of reality to add to our concept, one that is process based, creative and socio-environmental – what are the actions needed to create a square? – and which might help us understand this concept more deeply, through having this alternative means of expression.

In 1972, as Papert described early work with Logo, he claims, "The most important (and surely controversial) component of this impact is on the child's ability to articulate the working of his own mind and particularly the interaction between himself and reality in the course of learning and thinking" (Papert 1972, p. 246).

But would the process of drawing a square necessarily be enough in itself for the student to understand the complexity of such a concept or how it is learned? There are differing perspectives on how directed pedagogy for "ways of thinking" needs to be. Kalantzis, Cope, Chan and Dalley-Trim (2016, p. 454) claim, "Children in school are not taught the language of science explicitly. But by being exposed to the language of science while doing science, they learn its literacies, and so learn the ways of thinking embedded in those literacies". But Kozulin's (2003) Vygotskian perspective contradicts this view. Kozulin (2003, pp. 23–24) suggests that there may be difficulties in teaching particular ways of thinking or meta-cognition, as they are not a naturally occurring phenomenon and must be taught: "The acquisition of symbolic relationships requires guided experience; it does not appear spontaneously". It is interesting to note another potential reason for the lack of impact of the Logo project here. Kozulin claims that:

> Symbols may remain useless unless their meaning as cognitive tools is properly mediated to the child. The mere availability of signs or texts does not

imply that they will be used by students as psychological tools. This fact becomes particularly clear in the studies on the outcome of literacy ... Moreover, even literacy acquired in the nominally formal educational setting does not necessarily lead to the cognitive changes unless this literacy is mediated to a student as a cognitive tool.

(ibid, pp. 24–25)

Unless teachers "properly mediate" the new ways of thinking that use of a tool such as Logo or Scratch or MLfK offers, there is no guarantee that the students will recognise that what they are doing can provide them with a new way of thinking. This can be seen in the use of the Scratch tool, where it is perfectly possible for students to design and create fantastic animations without recognising the computational notions they are employing to do so, never mind the way in which representing those concepts computationally can help us understand how we learn about and develop our conceptual understanding of the world.

This is where Papert's teaching projects offered a powerful tool. He describes not only how these approaches encouraged learners to break down entities, activities or concepts being modelled into the processes that form them but also changed the practices of learning, offering opportunities for students to be "criticised on some other basis than 'right or wrong'" (Papert, 1972). In testing the merits of each other's teaching programs, Papert observed pupils advising, "Don't just tell him the right answer if he's wrong, give him useful advice". And discussing "what kind of advice is useful" leads deep into understanding "both the concept being taught and the processes of teaching and learning" (ibid).

AI, as a field exploring the modelling of human intelligence, is inherently explicit about questions of what thought is, what intelligence and learning are and how we might model the processes or *how to* of these concepts. Topics such as machine learning offer the opportunity to highlight what human ways of thinking and learning are and are not and the limitations of associated "mind-as-machine" metaphors of learning.

Let's take the example of the MLfK "Make me Happy" activity (see Figure 2.2). In addition to the computational concepts underpinning the Scratch aspects of the project and technical limitations such as quantity of training data examples, students also need to take into account and understand how words represent meaning, interpretation of emotions, social interaction and so on. In a sense, in this example task, the programme "learns from experience" and also from the student programmer, who can see themselves as "its teacher" or creator: what constitutes as "kind"? Who gets to decide which words are classified in particular ways, and which words represent particular categories? Discussion can arise as students more or less agree on what phrases may classify as "kind" and "unkind" and the binary distinction between such terms, necessary for computational modelling, offers opportunities to discuss the limitations of computation. Following Papert's

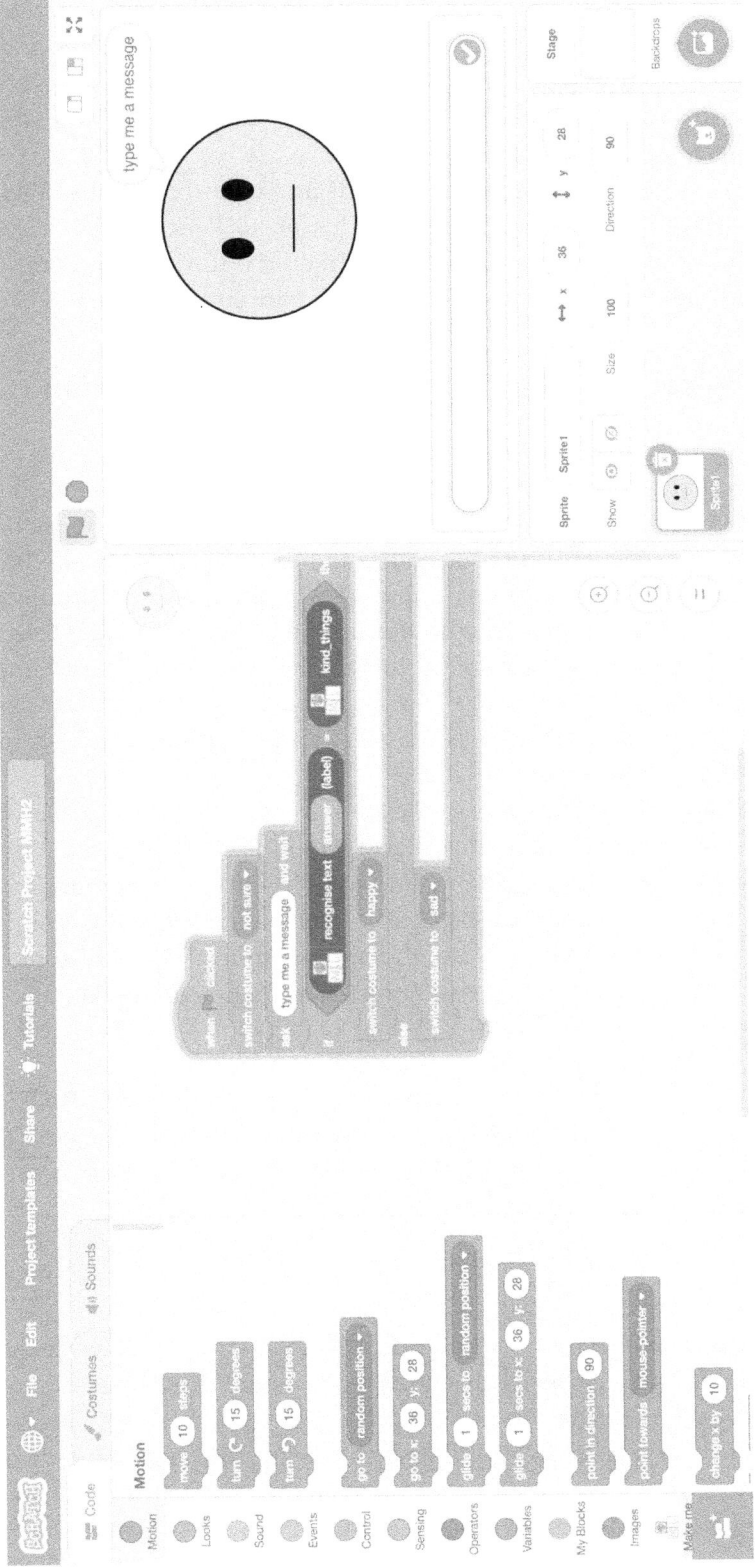

Figure 2.2 Screenshot of the Make me Happy activity, machine learning for kids.

suggestion of activities to write "teaching programs", the MLfK Chatbot activity develops a teaching support feedback chatbot, which can enable thinking and discussion about what "feedback" means within the learning process. Not only does this offer students insights into their own learning, but it also allows educators to see how learners perceive the art of teaching.

In this kind of activity, computational skills and concepts can be taught in harmony with developing a critical stance and, true to Papert's vision supporting learners to understand their own learning – but only if the activities are supported by discussions that make such perspectives and concepts explicit.

Epistemological pluralism – "ways in" to computation

In this process of representing concepts computationally, there is the possibility not only to learn about the concept as we understand it now and add to our individual complex of experiences and meanings for it but also to extend social, shared understandings of a concept and thus create new knowledge. In *An Exploration in the Space of Mathematics Education*, Papert (1996) describes how through viewing a well-known Euclidean theorem with children who were learning how to move the Logo turtle about in physical space, he was able to develop a new and shorter proof. This is at the heart of Papert's constructionism – a truly creative making of new knowledge and understandings, not just an individualistic construction of "already existing" knowledge.

However, this was not how these practices were, and arguably still are, positioned in the formal curriculum. Turkle and Papert (1990) argued that despite offering this potential for developing new forms of knowledge and understanding, the way in which computing education was implemented in schools and the way the practice was positioned had in fact the opposite effect. They describe how the notion of logic has been elevated above other forms of thought, but ought to be seen as simply one tool:

> ... "Hard thinking" has been used to define logical thinking. And logical thinking has been given a privileged status that can be challenged only by developing a respectful understanding of other styles where logic is seen as a powerful instrument of thought but not as the "law of thought." In this view, "logic is on tap, not on top."
>
> (ibid, p. 5)

Crucial to this perspective is recognition that each style of thinking has its limits (ibid) as do the pedagogies with which learners are introduced to computing practices. In Chapter 3, Chesterman describes Turkle and Papert's research with learners who feel they have to turn themselves into a different kind of person to succeed in computing. They describe this as stemming from an attitude to teaching computing, which does not allow for a plurality of "ways in" to the subject. This is a wasted opportunity as although, "... there has been a systematic construction of the computer as the ultimate embodiment of the abstract and formal", in fact, the practices of computing offer a platform for pluralism: it offers new concrete "ways in" to abstract and formal concepts. While a computer can provide a "virtual" model of reality, such as a flight simulator, it can equally provide a "real" model of a totally abstract idea, such as a design of a new building. It can act in the poised realm (Kauffman, 2016) between actuality and possibility.

Papert also recognised broader issues of inclusivity in computing education, although this was less of an explicit focus in his work. In *Mindstorms*, Papert suggested that whether children would learn to program or learn simply to use the computer as a pre-programmed tool would be determined by the different teaching environments in which they found themselves (Papert, 1980, pp. 29–30). Differences in pedagogic choices within these environments, based, for example, on socio-economic factors, could lead to the reinforcement of such inequalities.

Papert wrote extensively on the range of ways in which Logo was used by different teachers and different students in different contexts (see for example, Papert, 1985). He was adamant that there was no "right way" to use Logo, and any attempt to establish this would undermine the diversity of these different uses (ibid). But he did acknowledge there were wrong ways to use it, giving the example of a whole class of children writing a program to draw a house. He was concerned that in describing Logo and its activities, there was a tendency to focus on angular shapes, and when children were asked to "draw a house", the results were homogeneous, reinforcing a stereotype of the concept of "house". But Logo also provided ways to challenge that stereotype, and he describes an occasion where a child develops "smoky programming" in the attempt to include smoke in their house drawing.

Again, the importance here is allowing for difference in the outcome. When asked if children should have multiple "ways in" to mathematics, Papert points out that even the form of this question presumes a common end. He answers:

> My provocative conclusion will be to sketch a world in which we recognized the right to difference in the end as well as in the means. The therapeutic approach of the maths-anxiety specialists would be reversed: it is not neurotic but rather a highly coherent act of self-expression to resist learning in a way that goes against the grain of the deepest structures in one's own self.
> (Papert, 1986)

Papert's positioning challenges the distinction between "teacher" and "learner". Papert describes a teacher who was focused on how much her students would love an activity. He argues that while this is, of course, no bad thing, the teacher should not only focus on how the children feel, as this denies "the experience of self-learning, of letting yourself become totally engrossed in a personal learning experience, and through that, of recapturing the pleasure and excitement of learning something so naturally, in such an immersed and engrossed way, that you hardly notice that you're learning something" (Papert, 1985). The teacher's learning, about the concepts being explored and about how their students learn, became instrumental towards a fixed end goal, and Papert suggests that this is depriving the children also, "for the most valuable thing she could give them was an attitude and a feeling not so much about Logo as about learning, and surely she couldn't give this without letting herself experience it as well."

AI betwixt and between – learning about learning

While the socio-political conditions at the time of Logo's release did not support the take up of this tool in the ways that Papert had hoped, the increasing use of AI tools, both tangible and theoretical, in learning about computation has the potential to move us towards his vision. Kozulin's work would show that, in addition to Papert's approach, the processes of learning involved in those explorations needs to be explicitly highlighted. Don't create a chatbot to learn how to make a chatbot, but to explore understandings of language, communication and categorisation, to think about how we are learning through this making process.

AI research is constantly pushing the boundaries of how we understand our own learning and with each new discovery, as we explore this, there is then more to know. As Minsky suggests: "Whatever we learn, there is always more to learn – about how to use what was already learned." (Minsky, 1986, p. 100). Interest in robots for educational contexts tends to come from two different perspectives: how robots might support the learning of students in acting as teachers and how programming robots might help students learn about computing. This is not necessarily a helpful distinction as it arises from the existence of computing as a distinct subject with ends that are separate from other subjects. Robots have been increasingly used in classrooms where children might, for example, program Edbots to perform dance moves. But this still has a fixed and given end that serves no other purpose than getting a robot to perform dance moves and is another example of a "chocolate coated broccoli" activity. However, resituating the conversation and questioning around such an activity through cross-curricular links could make such an activity more meaningful. This might include the ways in which this kind of programming relates to cutting-edge robotics research and the distinction between AI and machine learning (for a clear discussion of these distinctions, see the MLfK book, Lane, 2021).

Developmental robotics, sometimes referred to as evolutionary robotics, is a field that develops cognitive architectures to sit within the body of a robot. Cangelosi defines the field as: "the interdisciplinary approach to the autonomous design of behavioral and cognitive capabilities in artificial agents (robots) that takes direct inspiration from the developmental principles and mechanisms observed in the natural cognitive systems of children". Research in this area challenges divisions between mind and body, nature and nurture, as researchers attempt to model how,

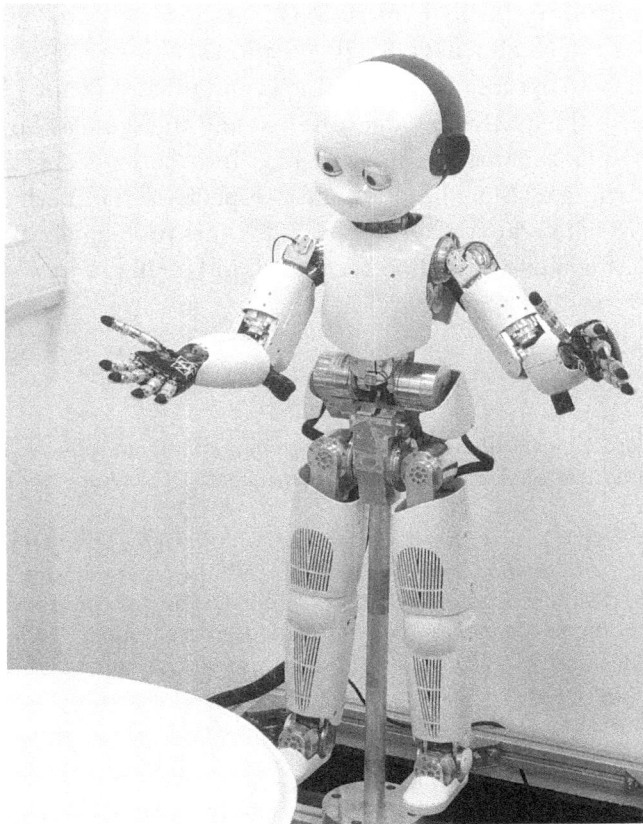

Figure 2.3 The iCub at the University of Manchester developmental robotics lab. Photograph by the author.

through the relationship of the "mind" and "body" (see, for example, Figure 2.3) of a robot, a robot can learn to associate a received image of an object with a sound that represents a word (see for example, Cangelosi and Stramandinoli, 2018; Štepánová, Klein, Cangelosi and Vavrečka, 2018). Here, robotics is trying to model the way humans bridge concrete and formal in the concept development process described above. Current research is also exploring the ways in which a child's interaction with a robot designed to be a tutor is similar, different or helpful in comparison to interactions with a robot designed to learn as a peer.

While classrooms aren't yet going to have a complex robot or the means to build cognitive architectures, there are small teaching tasks that can combine the "embodied" basic programming of an Edbot and the machine learning programming with MLfK. For example, programming MLfK to recognise confusion in a student, either through language or image recognition and an Edbot to perform a "sympathetic gesture", raises a host of possible explorations and discovery around questions such as: what is an emotion? What is "instinct"? How do these concepts develop in experience? Are we giving "instincts" to the robot and does the robot bear more responsibility for its own actions if its instincts are "given"? What responsibility does the creator have for its creation's actions? Perhaps most importantly, we can start to explore what learning is and what it is not – how is the way this robot, or our trained MLfK model different and the same as how we learn?

Embedding Papert's pedagogical principles into both teachers' and students' teaching and learning of AI techniques and perspectives will perhaps enable computing to achieve what Papert saw in Logo: "it's this protean ability to take different forms – and, if you use it right, to become a kind of mirror in which you can see reflections of yourself" (Papert, 1985).

References

Agalianos, A., Noss, R. and Whitty, G. (2001). Logo in mainstream schools: The struggle over the soul of an educational innovation. *British Journal of Sociology of Education*, 22(4), pp.479–500.

Banks Gatenby, A. (2017a). *Developing Perspectives of Knowledgeability through a Pedagogy of Expressibility with the Raspberry Pi* (Ph.D. Dissertation). Manchester, United Kingdom.

Banks Gatenby, A. (2017b). Developing critical understanding of computing with the Raspberry Pi. *International Journal of People-Oriented Programming*, 6, (2), pp.1–19.

Brennan, K. and Resnick, M. (2012). *New Frameworks for Studying and Assessing the Development of Computational Thinking*. Proceedings of the 2012 Annual Meeting of the American Educational Research Association, Vol. 1, Vancouver, April 13–17, 2012, p.25.

Bullen, M. (1998). Participation and critical thinking in online university distance education. *Journal of Distance Education*, 13(2), pp.1–32.

Cangelosi, A. and Schlesinger, M. (2018). From babies to robots: The contribution of developmental robotics to developmental psychology. *Child Development Perspectives*, 12, pp.183–188.

Cangelosi, A. and Stramandinoli, F. (2018). A review of abstract concept learning in embodied agents and robots. *Philosophical Transactions of the Royal Society Biology*, 373(1752). http://doi.org/10.1098/rstb.2017.0131

Chan, N., Ho, I. T. and Ku, K. Y. L. (2011). Epistemic beliefs and critical thinking of Chinese students. *Learning and Individual Differences*, 21(1), pp.67–77.

Kalantzis, M., Cope, B., Chan, E. and Dalley-Trim, L. (2016). *Literacies*. 2nd edn. New York: Cambridge University Press.

Kauffman, S. (2016). Answering Descartes: Beyond Turing. In Barry Cooper, S. and Hodges, A. (eds) *The Once and Future Turing: Computing the World*. Cambridge University Press. St Ives.

Kozulin, A. (2003). Psychological Tools and Mediated Learning. In Kozulin, A., Gindis, B., Ageyev, V. S. and Miller, S. M. (eds) *Vygotsky's in Educational Theory in Cultural Context*. New York: Cambridge University Press, pp.15–38.

Lane, D. (2021). *Machine Learning for Kids: A Project-Based Introduction to Artificial Intelligence*. San Francisco: No Starch Press.

Luckin, R. (2018). *Machine Learning and Human Intelligence the Future of Education for the 21st Century*. London: UCL Institute of Education Press (University College London Institute of Education Press).

Minsky, M. (1986). *The Society of Mind*. New York: Simon and Schuster.

Minsky, M. and Papert, S. (1972). Artificial Intelligence Progress Report. Retrieved 20.01.22 from: http://dailypapert.com/wp-content/uploads/2020/07/artificial-intelligence-progress-report-Papert-Minsky-AIM-252.pdf

Mitchell, W. J. (1996). *City of Bits: Space, Place, and the Infobahn*. Massachusetts: MIT Press.

Noss, R and Hoyle, C. (2013). Richard Noss and Celia Hoyles on Seymour Papert. Retrieved 20.01.22 from: https://www.youtube.com/watch?v=UrYekbT6X8w

Papert, S. (1972) Teaching children thinking. *Programmed Learning and Educational Technology*, 9(5), pp.245–255.

Papert, S. (1980). *Mindstorms: Children, Computers and Powerful Ideas*. New York: Basic Books.

Papert, S. (1982) Tomorrow's Classrooms? *Times Educational Supplement* March 5, 1982, pp.31–32, 41. Retrieved 20.01.2022 from: http://dailypapert.com/tomorrows-classrooms/

Papert, S. (1985). Different visions of logo. *Computers in the Schools*, 2(2–3), pp.3–8

Papert, S. (1986). Beyond the Cognitive: The Other Face of Mathematics. In *Proceedings of the International Conference for the Psychology of Mathematics Education*. Retrieved 20.01.2022 from: http://dailypapert.com/wp-content/uploads/2015/07/BeyondTheCognitive.pdf

Papert, S. (1996). An exploration in the space of mathematics educations. *International Journal of Computers for Mathematical Learning*, 1(1), pp.95–123.

Papert, S. (2000). What's the big idea? Toward a pedagogy of idea power. *IBM Systems Journal*, 39(3), pp.720–729.

Papert, S. (2005). You can't think about thinking without thinking about thinking about something. *Contemporary Issues in Technology and Teacher Education*, 5(3/4), pp.366–367.

Sentance, S. and Csizmadia, A. (2017). Computing in the curriculum: challenges and strategies from a teacher's perspective. *Education and Information Technologies*, 22, pp.469–495. https://doi.org/10.1007/s10639-016-9482-0

Štepánová, K., Klein, F. B. Cangelosi, A. and Vavrečka, M. (2018). Mapping language to vision in a real-world robotic scenario. *IEEE Transactions on Cognitive and Developmental Systems*, 10(3), pp.784–794.

The Institute for Ethical AI in Education (2021). The Ethical Framework for AI in Education [Online]. Retrieved from: https://fb77c667c4d6e21c1e06.b-cdn.net/wp-content/uploads/2021/03/The-Ethical-Framework-for-AI-in-Education-Institute-for-Ethical-AI-in-Education-Final-Report.pdf

Turkle, S. and Papert, S. (1990). Epistemological pluralism and the re-evaluation of the concrete. *SIGNS: Journal of Women in Culture and Society*, 16(1), pp.128–157.

Zook, M. A. and Graham, M. (2007). Mapping DigiPlace: Geocoded internet data and the representation of place. *Environment and Planning B: Planning and Design*, 34(3), pp.466–482. https://doi.org/10.1068/b3311

3 Keeping it real

Helping learners navigate the concrete and abstract

Mick Chesterman

Introduction

In the UK, computing and computational devices are all around us. Young people interact with them in many ways including general communication, games, social media and creative apps. Computing as a subject and coding as an activity can draw on the diverse ways that computing touches people's concrete lives. A quick search of the web for creative computing or tech for kids yields a multitude of activities, devices and materials designed to engage the hobby interests of young people. Physical examples include fashion- and textiles-based computing, robotics, colourful lighting displays and programming Lego constructions. This chapter celebrates the value and fun involved in coding in these contexts. However, while the range of materials and possible creative projects is promising, aligning creative opportunities with an exam-assessed curriculum is challenging.

This chapter highlights the value and challenges of implementing hands-on teaching approaches in a UK schooling context. It begins by looking at interpretations of inclusion and specifically the Universal Design for Learning (UDL) framework. The terms concrete and abstract and their relevance to coding and definitions of computational thinking (CT) are explored. The second half of the chapter then turns to practical ways teachers can help learners navigate abstract concepts and benefit from hands-on experience of coding. To do this, it outlines some techniques promoted by the National Centre for Computing Education (NCCE). Throughout this part of the chapter, there is a focus on linking these educational practices with inclusive approaches and learner engagement.

Inclusion and inclusive pedagogies

The term inclusion in education no longer addresses solely children with special educational needs and disabilities (SEND) but also examines barriers of culture and other exclusionary elements of the school environment and discourse

(Black-Hawkins et al., 2008). The issue of alienation from the culture of computing in schools has been identified as a concern, especially for girls and some ethnic minorities (The Royal Society, 2017). To be inclusive, schools and teachers must identify both traditional SEND issues and wider cultural barriers to participation in computing classrooms and help students overcome them. Beyond the important technical accessibility tools and assistive learning technologies – for example, text in different sizes and screen readers – a diversity of inclusive teaching strategies is also needed.

One way to address SEND issues is to use differentiation to adapt the standard lesson plan for learners needing special support. However, this view of a standard, optimal learner pathway is not supported by recent research in neurodiversity, which suggests there is no one optimal way for students to learn. Inclusive pedagogies take a different approach to differentiation, which places more power in the hands of learners to choose the path that is most appropriate for them. All students are given a greater choice of materials and activities from the start suiting the varied needs of all students. This has the benefit of removing the stigmatisation of some learners having to undertake work that is perceived as being created for *low-achieving* students. These principles – among others – are presented in a framework called UDL.

Key concept – UDL

UDL is a set of design principles aimed at educators to help them design learning experiences that incorporate diverse ways to engage pupils and to represent the concepts being communicated. The UDL framework provides guidelines for three key areas of representation, engagement and expression/action (CAST, n.d.). You can use the following summary of UDL principles as a checklist to help you to plan inclusive teaching activities. A more comprehensive outline is provided by the Center for Applied Special Technology organisation.[1]

Multiple means of representation: Are you presenting material and concepts in multiple formats? For example, spoken presentations, written documents, graphics, hands-on activities or audio material. Are you clarifying new language and symbols in diverse ways? Are you providing important background knowledge and highlighting important patterns and relationships in the knowledge you explore?

Multiple means of engagement: Have you been able to increase student choice and the relevance of your material to spark learners' excitement? Are you using a variety of ways to allow your learners to focus on their goals, maintain self-belief and sustain their effort as individuals and in group work?

Multiple means of expression/action: No one particular action or format of expression will be best for all students. Are you able to offer a choice in how your students interact with materials and tools (particularly assistive technologies) and allow students choice in the media they use for communication and construction? How much are students able to set their own learning goals and monitor their own progress?

UDL places great value on the personal relevance, choice and authenticity of learning experiences. As a way to encourage engagement, UDL suggests setting choices of concrete learning goals that are relevant to the learners. This learner-led approach is very different from a traditional instruction-based, directive approach to teaching. The diverse learning pathways offered can be unfamiliar for both teachers and students. The UDL guidelines recognise this and provide information to support teachers to implement them. As educators, we may need to build our own abilities and familiarity with learner-led approaches as well as grow the autonomy of our students.

One area of UDL that teachers can implement straightforwardly is to represent concepts in the classroom in a diversity of ways. In a related study, researchers Cook et al. (2016) explored the alignment of UDL with another framework, CRA, which consists of a three-stage model to support learners to develop concepts (Fyfe et al., 2014). The researchers outline how the three stages of CRA (concrete, representational and abstract) align with key UDL principles, most specifically multiple ways to represent knowledge to aid learner perception and comprehension. In short, first teachers introduce a physical, concrete model of the concept, then progress to iconic forms, for example, graphics or pictures; finally, learners work with more abstract models of the concept. The CRA framework is an example of concept popular in mathematics research and practice called concreteness fading where concepts are introduced in concrete examples and then learners are supported to understand and represent them in more abstract ways.

When reading about different approaches to teaching computing, the terms concrete and abstract are used commonly. For example, the concrete practice of coding is a good way for learners to work with more abstract computing concepts. The following section explores the utility of these terms to explore inclusive approaches to teaching, especially in relation to an understanding of CT.

Computational thinking: The abstract and concrete

Concrete and abstract learning approaches

In an everyday sense, concrete objects are ones you can get hold of and abstract objects exist only as concepts. A pound coin is concrete, but the idea of profit is abstract. In everyday usage, abstract knowledge may be harder to grasp than more concrete understandings. For example, we might ask for a concrete example if we don't understand a more abstract definition. Here, it is the use of something in a context that makes something concrete. In traditional conceptions of education, abstract knowledge is often perceived to be of greater value. If you can understand a concept as it applies in different situations, then this ability to transfer it and have a more global understanding is held as a higher form of knowledge. This concept is popular in education in many forms, for example, Piaget's influential model of developmental stages, specifically in the progression to more abstract thinking in the transition from the concrete (operational) to formal (abstract) stages (Burman, 2021).

Not all educators agree with the supremacy of more abstract ideas of knowledge and are keen to celebrate the value of concrete exploration by learners. Seymour Papert and Sherry Turkle's work on creative computing at Massachusetts Institute of Technology created a legacy that includes the development of the Scratch programming tool and the use of physical computing in education. Papert and Turkle (1990) thought it was vital that we value and recognise concrete approaches to computing and coding. While they do not discard the technical value of abstract approaches, they draw on feminist theory to make a convincing case that approaches like abstract planning and formal language can be off-putting to certain learners and especially girls. They argue that the process of finding solutions to coding issues for novice coders should be a matter of personal preference. As well as the terms abstract and concrete – they use the terms top-down and bottom-up approaches to learning. In a bottom-up approach, problems are tackled piece by piece, experimentally. Desmond Tutu once said that "there is only one way to eat an elephant: a bite at a time". Bottom-up (concrete) coders take this approach.

To clarify this, Papert and Turkle give the example of a young coder Lisa, who is aware of a more formal way to approach the kinds of programming tasks she is undertaking, but maintains that way doesn't work for her. As she continues her journey as a coder, this frequent message that she is doing things the wrong way demotivates her enthusiasm around coding. Abstract coding concepts are tools for thinking. But they are only useful to the learner if they match with her experiences. Forcing the learner to adopt an abstract approach in this example is counterproductive as it undermines her experience and progress. The authors outline that the danger of prioritising teaching and testing of abstract concepts is to devalue this bottom-up approach to coding. This way of coding has also been called a craft approach. It is a way of doing things that has been shown to have a lot of value in many professions.

Thus as inclusive educators, we have a duty to allow learners to follow a learning path that suits them as much as is practical within the constraints of the curriculum. A conceptual tool that may help teachers to help learners to navigate between the abstract and concrete is semantic profiling.

Key concept – semantic profiles

Semantic profiles chart the use of more concrete (high semantic gravity) language and more abstract (high semantic density) concepts and patterns as they emerge in classroom situations (Macnaught et al., 2013). Exploring semantic profiles is being promoted by NCCE as an aid to teachers wanting to plan their lessons in a way that communicates the key abstract curriculum knowledge that students will need for exams, and to also allow them to put the concepts into practice to build real coding skills and to make valuable connections to personal experience. A Quick Read on semantic profiles is available on the NCCE website.[2]

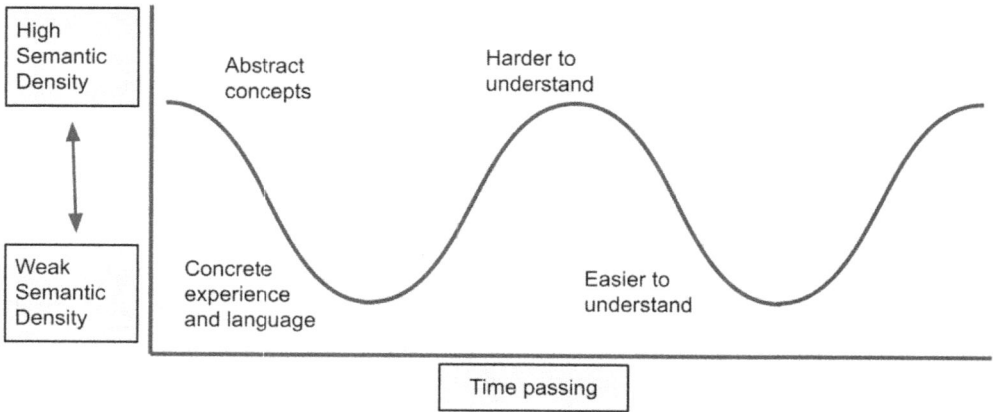

Figure 3.1 A semantic profile with semantic waves.

Research carried out by Curzon et al. (2020) in a computing education context outlines the value of semantic profiles in wave shape as opposed to a flatline, which remains too much in concrete examples or more abstract concepts. This research highlights the value of unpacking, exploring and then repacking ideas during the course of a lesson. A student's understanding of a concept may deepen a little bit each time it is applied in practice and then reconnected with the abstract.

Semantic waves

Examples showing semantic wave profiles usually start with the abstract or with high semantic density. See Figure 3.1 for an example. The advice of starting with more abstract terminology and concepts may seem to be in opposition to the approach of concreteness fading as explained with the CRA method outlined above. However, on examination of the research example carried out in the research above, the process of starting with concepts may only involve giving a short outline of the concept that is being explored and saying that this will be illustrated in a following concrete activity.

Activity – recognising and exploring the concrete and abstract in your teaching

To help you balance your lessons, it is useful to identify which parts address more abstract conceptions of computing knowledge and those that are more concrete. To do this, ask yourself the following questions about your lessons:

- What are the core skills and concepts I want to communicate in my lesson? How many of them are more abstract and how many are more concrete? Do I have a good balance?

- How do learners first meet core concepts? How are they re-enforced? Is it through an abstract definition, supported by a more concrete explained example, or via hands-on work that students are engaged in?
- What kind of semantic profile do your lessons follow? Are you able to avoid "flatlining", where students explore only abstract or concrete language?
- Are you alternating between concrete and abstract poles in a way that helps deepen the understandings of students as they link curriculum concepts to practical experience by helping them link curriculum concepts to practical experience and back again?

Definitions of computational thinking

The promotion of CT has been a key factor in the development of the UK's computing curriculum. However, the claims of early advocates that CT skills could be applied widely in subjects beyond computing are now advanced more cautiously to avoid the danger of over-promising (Tedre and Denning, 2016). We can use the distinction between concrete and abstract to examine the differences between two popular interpretations of CT. The first is an influential take from Jeanette Wing. "The most important and high-level thought process in computational thinking is the abstraction process. Abstraction is used in defining patterns, generalizing from instances, and parameterization" (Wing, 2011). Many learning resources designed to support the computing curriculum present this principle as four key pillars of CT: decomposition, pattern recognition, abstraction and algorithmic thinking (BBC Bitesize, n.d.). The essence here is to deal with concepts and principles as abstract and separate from the context of coding.

Another widely used definition of CT by Brennan and Resnick (2012) was developed in response to a thought experiment "How do we describe what Tim, Shannon, and Renita are learning as they participate as designers of interactive media with Scratch?". The researchers took a grounded/situated approach to mapping the potential learning dimensions of students designing and coding collaborative, creative computing projects. The resulting map they created includes computational concepts, computational practices and computational perspectives.

- *Computational concepts* include sequences, loops, parallelism, events, conditionals, operators and data, thus representing the mechanics of coding structures.
- *Computational practices* include debugging, iteration, reuse and remixing, abstracting and taking a modular approach.
- *Computational perspectives* such as expressing, questioning and connecting were observed in the behaviour of learners completing their coding designs.

This interpretation of CT, based on observation of learners in action, is more accessible to teachers and learners in comparison with the more abstract interpretations of CT as they can easily recognise their own practice. To give a specific example, rather than decomposition, the applied framework outlines taking an iterative, incremental approach to problem solving and arranging code in modules.

This broader, process-driven definition of CT has been used and adapted by many organisations seeking to support the new computing curriculum. As such, it may be familiar from websites, posters and other supporting material created by groups like Barefoot Computing. Lye's extensive review of teaching CT (Lye and Koh, 2014) used Resnick and Brennan's definition as the basis for the review, which points to the widespread use of this more applied approach. The wider definition of CT here assumes an environment where learners are engaged in the collaborative coding of a computing project. The review above and the influential framework used by computing at school (Csizmadia et al., 2015) have included elements of this applied framework as well as other more abstract CT concepts.

The popularity of the CT frameworks describing concrete and collaborative learning help support the development of inclusive teaching approaches. However, the tension between abstract and, thus, potentially transferable knowledge and context-bound concrete knowledge exists not only in definitions of CT and but also in approaches to teaching and assessment. In the next section, I will explore this tension and look at recent responses in the field of UK computing education.

Models for teaching computing in the classroom

Classroom practice is strongly influenced by curriculum content and more specifically the format of exam questions. The removal of coursework from General Certificate of Secondary Education (GCSE) exams due to the widespread sharing of worked examples on-line created a real challenge for those setting exams. Exam boards had to rethink how to test the practical programming experience and ability of students in a written exam setting. At GCSE level, most exam questions test the more applied definitions of CT, particularly the process of writing, analysing and revising algorithms in the form of written code examples. Students are required to demonstrate and explain fundamental code building blocks and approaches from first principles. However, exam questions on coding are by nature fragmented and decontextualised compared to practical project coding experience. The written exam questions presenting small coding challenges are necessarily fragmented in order to test a particular part of the curriculum. A focus on written exam approaches in the classroom can reduce time to explore more authentic projects where students follow their own interests and develop skills in design, debugging and other troubleshooting practices.

Spending a lot of time in lessons addressing decontextualised exam questions has potentially negative impacts on student engagement and such disengagement is a serious barrier to inclusion (Kanevsky and Keighley, 2003). In an ideal situation, practical work can deepen and broaden more abstract knowledge. However, given the reality of schools' focus on exam results, the teacher must balance the development of students' hands-on coding practices with the ability to recognise and respond to more abstract written questions. Having proficiency in a variety of relevant teaching approaches can be helpful to resolve this tension. The NCCE has produced a series of resources based on research on computing practices to help teachers. They have provided a set of 12 principles for teaching computing[3] and supporting Pedagogy Quick Read documents[4] aimed at teachers to explain and promote key research-informed teaching techniques. The following sections summarise some of these principles.

PRIMM: PRIMM stands for predict, run, investigate, modify and make. This model helps learners adopt coding practices and computational concepts through providing a concrete code example that they run after predicting what it does. Learners then make changes to the existing code before finally creating code from scratch. Asking students to identify target computing concepts in code examples allows teachers to guide students towards key CT processes or algorithmic details. Thus, PRIMM is well suited to the classroom as starting with the prediction of code allows a whole class of learners to unpack and repack the same set of concepts in a restricted time scale. This process enables students to more easily tackle the formal problem solving, paper-based questions of the GCSE exams. The use of code examples and a structured set of varied activities aligns well with UDL principle of representing knowledge in a variety of means. For a more detailed summary of the PRIMM approach, see the Quick Read pedagogy article.[5]

Unplugged activities: Unplugged activities are carried out away from the computer and aim to illustrate computing concepts. Unplugged activities often use very familiar non-school examples and draw on learners' understanding of their own concrete experiences. As an example, teachers may use cooking recipes as a way to illustrate the importance of correct sequencing in a code context. Unplugged activities are also often embodied activities. Embodied ways of learning involve moving beyond a computational view of cognition to recognise the importance of physical experience and emotions in the learning process (Settoducato, 2017). Thus, the practice is very much in line with the UDL principle of allowing multiple forms of expression and action for learners and highlighting patterns of knowledge representation. To help learners integrate abstract concepts, teachers should link unplugged activities with concrete coding activities. A semantic wave approach as outlined above can help teachers explore this process.

Pair programming: Pair programming groups students in pairs and divides coding into two roles. One student undertakes hands-on coding while the other is free to think about the abstract design of the overall program. A benefit of pair programming is to increase coding confidence as students build their experience of the different roles involved. To help novice coders, teachers should model and break down the processes involved. Pair programming involves social learning elements and can model greater choices for students in the ways they solve problems. The process of building an identity in a community with the help of peers is key to a socio-cultural understanding of how learners pick up coding in a classroom (or other settings). The importance of a coding community is explored in this collection on design and project approaches. A summary of pair programming roles and tips on how teachers can help learners to adopt them in present in a Quick Read document from NCCE.[6]

Observing and assessing hands-on practices: One way to address the tension between giving learners the freedom to pursue their own concrete coding goals as is the use of observational techniques to assess user progress. Skilled observation allows teachers to catch learners when they are engaged with their own personally meaningful project and their motivation to solve problems is high. The NCCE has created a Quick Read on observation.[7] A summary of that document follows:

- *Structured observations:* Before lessons, teachers create a framework of the behaviour or use of concepts they want to observe during interactions with students, via recordings or created work.

- *Unstructured observations:* Teachers record or reflect on some of the more unexpected turns that happen during the lesson, often after the event from memory or recordings.

- *Verbal protocols:* Teachers assess the learning and understandings of students by asking them to talk aloud the way they are solving problems and undertaking tasks.

This kind of observation is time-consuming and can benefit from extra classroom support. However, this effort can be justified as an inclusive measure to support a diversity of learner expression and assessment recommended by the UDL guidelines.

The use of concept maps and learning frameworks: To facilitate structured observation, a predetermined framework of the kind of behaviours, practices and concepts that align with the project work being undertaken is extremely useful. One way that teachers can develop such frameworks is via a technique called concept maps. The NCCE has created a Quick Read guide for teachers to create concept maps, which focuses on more technical knowledge[8]. In this document, concept maps are presented as a way for teachers to model and for students to explore connections between concepts. They can also be used as a map or a matrix to help observation. Providing students with this framework can also help them navigate their learning journey. Having potential learning clearly mapped out and involving students in self-monitoring increases the efficiency of the observation process. The visual nature of the maps also aligns with UDL guidelines on presenting concepts via multiple means. I explore project approaches in Chapter 8 of this book.

Conclusion

The purpose of this chapter has been to celebrate the educational value of hands-on, concrete coding as an inclusive way to explore the computing curriculum. As computing educators, we are lucky to have rich and engaging resources at our disposal, which invite tinkering and learning through trial and error experimentation. This chapter has explored the value of UDL principles for teachers looking to inclusive practice and the possibility to assess content knowledge through observing students during hands-on work. We have seen researchers Papert and Turkle celebrate the value of concrete at a very early stage of computing education. More recently, Resnick and Rusk (2020) from the Scratch research community caution against recent tendencies to adopt predominately formal approaches to computing including: too much memorisation of computational terms rather than application, devaluing hands-on coding compared to abstract concepts, not enough time devoted to complete projects and, finally, learners not given enough choice in their coding projects. I share the concern of these researchers that the potential for computing to build creative and design thinking competencies is not being optimised in formal education.[9] Teachers in the UK continue to experience tensions between encouraging experimentation and the pressure to bring students' attention back to underlying concepts that are assessed through more abstract test material. The NCCE is playing a pivotal role in supporting teachers to recognise and navigate the abstract and concrete via teaching techniques. I explore the principles of design- and project-based approaches, which also have the potential to aid teachers to balance the requirements of the curriculum and the value of letting students spend more getting their hands dirty in the concrete of coding.

Notes

1. https://udlguidelines.cast.org/
2. https://blog.teachcomputing.org/quick-read-6-semantic-waves/
3. https://blog.teachcomputing.org/how-we-teach-computing/
4. https://blog.teachcomputing.org/tag/quickread/
5. https://blog.teachcomputing.org/using-primm-to-structure-programming-lessons/
6. https://blog.teachcomputing.org/quick-read-pair-programming-supports-learners/
7. https://blog.teachcomputing.org/using-observation-techniques-to-record-student-behaviour-for-research-or-evaluation/
8. https://blog.teachcomputing.org/using-concept-maps-to-capture-communicate-construct-and-assess-knowledge/
9. https://web.media.mit.edu/~mres/papers/CACM-Coding-At-Crossroads.pdf

References

BBC Bitesize, n.d. Introduction to computational thinking – KS3 Computer Science Revision [WWW Document]. BBC Bitesize. URL https://www.bbc.co.uk/bitesize/guides/zp92mp3/revision/1 (accessed 3.26.21).

Black-Hawkins, K., Florian, L., Rouse, M., 2008. Achievement and inclusion in schools and classrooms: Participation and pedagogy, in: *Artículo Presentado En La Conferencia de British Educational Research Association*. Universidad Heriot Watt. Edinburgh.

Brennan, K., Resnick, M., 2012. New frameworks for studying and assessing the development of computational thinking. Paper presented at Annual American Educational Research Association Meeting, Vancouver, BC, Canada.

Burman, J.T., 2021. The genetic epistemology of Jean Piaget, in: *Oxford Research Encyclopedia of Psychology*. Oxford University Press. https://doi.org/10.1093/acrefore/9780190236557.013.521

CAST, n.d. About universal design for learning [WWW Document]. CAST. URL https://www.cast.org/impact/universal-design-for-learning-udl (accessed 2.13.22).

Cook, S., Rao, K., Cook, B., 2016. Using universal design for learning to personalize an evidence-based practice for students with disabilities, in M. Murphy, S. Redding, J. Twyman (Eds.), Handbook on personalized learning for states, districts, and schools (pp. 239–247). Temple University, Center on Innovations in Learning, Philadelphia, PA. Retrieved from www.centeril.org

Csizmadia, A., Curzon, P., Dorling, M., Humphreys, S., Ng, T., Selby, C., Woollard, J., 2015. Computational thinking – a guide for teachers. Computing at School.

Curzon, P., Waite, J., Maton, K., Donohue, J., 2020. Using semantic waves to analyse the effectiveness of unplugged computing activities, in: Proceedings of the 15th Workshop on Primary and Secondary Computing Education, WiPSCE '20. Association for Computing Machinery, New York, NY, pp. 1–10. https://doi.org/10.1145/3421590.3421606

Fyfe, E.R., McNeil, N.M., Son, J.Y., Goldstone, R.L., 2014. Concreteness fading in mathematics and science instruction: A systematic review. *Educational Psychology Review* 26, 9–25. https://doi.org/10.1007/s10648-014-9249-3

Kanevsky, L., Keighley, T., 2003. To produce or not to produce? Understanding boredom and the honor in underachievement. *Roeper Review* 26, 20–28. https://doi.org/10.1080/02783190309554235

Lye, S.Y., Koh, J.H.L., 2014. Review on teaching and learning of computational thinking through programming: What is next for K-12? *Computers in Human Behavior* 41, 51–61. https://doi.org/10.1016/j.chb.2014.09.012

Macnaught, L., Maton, K., Martin, J.R., Matruglio, E., 2013. Jointly constructing semantic waves: Implications for teacher training. *Linguistics and Education* 24, 50–63. https://doi.org/10.1016/j.linged.2012.11.008

Papert, S., Turkle, S., 1990. Epistemological pluralism and the revaluation of the concrete. *Signs*, 16(1), 128–157. http://www.jstor.org/stable/3174610

Resnick, M., Rusk, N., 2020. Coding at a crossroads. *Communications of the ACM* 63, 120–127. https://doi.org/10.1145/3375546

Settoducato, E., 2017. Of pedagogy and potentiality: Embodied learning and collaborative storytelling through pop-up exhibits. *Journal of New Librarianship* 2, 117–121. https://doi.org/10.21173/newlibs/3/5

Tedre, M., Denning, P.J., 2016. The long quest for computational thinking, in: Proceedings of the 16th Koli Calling International Conference on Computing Education Research. Presented at the Koli Calling 2016: 16th Koli Calling International Conference on Computing Education Research, ACM, Koli, Finland, pp. 120–129. https://doi.org/10.1145/2999541.2999542

The Royal Society, 2017. After the reboot: Computing education in UK schools. The Royal Society.

Wing, J., 2011. Research notebook: Computational thinking—What and why. *The Link Magazine* 6, 20–23.

4 AI is racist

Richard A. Dunk

Introduction

There's an oft-quoted definition of an algorithm as "a word used by programmers when they don't want to explain what they did". Although obviously tongue-in-cheek, this definition does allude to the mystique around algorithms as a "black box" whose inputs and outputs can be observed, but whose "internal" processes themselves are less transparent. This understanding of algorithms is somewhat misleading when we consider "hand coded" algorithms (such as the ubiquitous bubble sort or the Fizz-Buzz game that stereotypically appears in interviews for junior programmers) in which the internal workings are perfectly visible in the codebase. However, in the case of self-constructing algorithms, such as those produced by machine-learning (ML) processes, these interior workings can indeed be impenetrable.

In a broad sense, ML describes computational processes that use "experience to improve performance or to make accurate predictions" (Mohri et al., 2018, p. 1) and is one aspect of what is colloquially referred to as artificial intelligence (AI). A common example is a neural network, where a model is built and "trained" with existing data before being applied to new scenarios to make predictions and improve understanding. The resulting model can take inputs and provide outputs, but the intermediary steps are not "human readable", leading to an algorithmic system that is "multi-layered", as shown in Figure 4.1, where the inputs and outputs are "coupled" through a hidden, machine-generated, processing layer.

Given the meteoric rise of ML, examples of algorithms are easy to find. One simple example is Gautam (2020), who trained an ML model to differentiate between images from rugby or football games; the resulting algorithm was able to do so with 83% accuracy. A similar example on a larger scale is Google's *Quick, Draw!* site (Google, 2016), where quickly drawn "doodles" of a given object are algorithmically identified in real time. Not only does each drawing test the model by

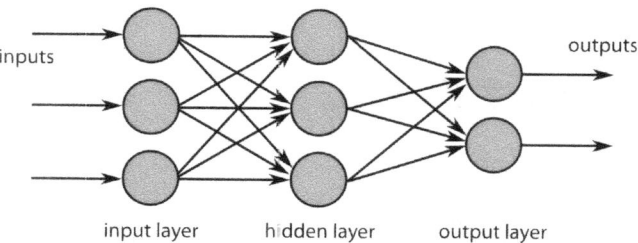

Figure 4.1 Schematic showing "layers" in a neural network (Image from WikiMedia Commons user Chrislb used here under the Creative Commons Attribution-ShareAlike 3.0 Unported license).

seeing if it can guess the correct output from a given input, but each drawing also constitutes a piece of categorised data that can be used to refine the model further. After nearly six years of collecting data, the model is highly refined; I found the speed at which the algorithm could identify my awful sketches disturbingly fast. You can try it yourself at https://quickdraw.withgoogle.com/. This demonstrates two further strengths of neural networks: the ability to quickly make digital inferences from vast sets of complex "real world" data, and the continual refinement of existing models by building feedback loops into the process to further improve the accuracy of outputs.

More serious examples of ML-developed algorithms can be found to demonstrate the potential for such models to positively benefit society, such as the team from The Massachusetts Institute of Technology's (MIT) Computer Science and Artificial Intelligence Laboratory that created a model trained to predict if a patient is likely to develop breast cancer in the future from a mammogram image (Yala et al., 2019), or the spate of companies analysing complex multivariate data sets to predict which utility infrastructure is most as the risk of failure (Waddell, 2019) to help allocate maintenance resources efficiently. However, with such examples, we can begin to imagine the ethical implications of living with AI decision-making and the complex relationship between lived human realities and the "black box" of machine-generated models. What are the consequences of a machine identifying you as having a high risk of breast cancer? How might infrastructure decisions made using ML models cause suffering for those whose areas aren't prioritised? And, importantly for this book, how might inclusive teaching practices and an increased focus on ethical computing help recognise the point at which "human intervention" become a necessity?

On a small scale, the ethical implications of AI decisions are fairly well considered. Applications such as automated facial recognition employed by police forces raise complex issues around privacy and rights during criminal processes (Purshouse and Campbell, 2019), and its controversial use in the Hong Kong protests of 2019 casts overtly Orwellian overtones (Doffman, 2019). Recent rethinkings of the eponymous "trolley dilemma" to account for the ways that a self-driving car should respond when "it has the choice between crashing into five persons on the right or, if it changes its trajectory, one person on the left" (Wolkenstein, 2018, p. 164) also highlights the urgent need for ethical computing in small-scale, personal situations. However, ethical considerations of ML algorithms for society more broadly are perhaps less widely (or at least less succinctly) considered.

The personal impact and rapid responsiveness of ML models that have been trained on large, proprietary datasets to provide "personalised" online experiences are clear to see; the music recommendations on my Spotify account are far more nursery rhyme and Disney-oriented since I became a parent, and automated playlists that sit Audioslave and Nirvana alongside the Paw Patrol theme tune produce a jarring juxtaposition that only a machine could think was a good idea! In a recent tweet, author A. H. Reaume attributes a similar experience directly to ML models:

> I had the perfect Instagram ad algorithm honed—all workout wear, and elaborate lacy lingerie.
>
> Then a friend of mine linked me to some blocks he's buying his kids and now all I am seeing are educational toys. Lol.
>
> The blocks were so cool! No regrets! Lol.
>
> But it's amazing just how much we are tracked online.
>
> (Reaume, 2020, reproduced with permission)

While ML models may just be a collection of self-compiling if statements, it's clear that such algorithmic ideas, and their social impact, have entered popular consciousness in a way that problematises computational thinking. Hence we, as educators in Computer Science, need to both raise awareness of the existence and social implications for such algorithms and work to reduce potentially negative impacts. It is clearly beyond the scope of this chapter to document these broad issues in their entirety, so I have centred the discussion around the ways that ML algorithms trained on big datasets are propagating and exacerbating inequalities seen in society more widely.

"Big data" and algorithmic abuse

As I have already described, ML algorithms are produced by "training" software to generate outputs based on existing data. Whether these "outputs" are automated image identification, detecting emotional sentiment in texts, or computerised trading systems controlling where our pensions are invested, the success of these algorithms is initially dependent upon existing data in contemporary ways that reflect the old adage "garbage in, garbage out". Bigger existing datasets mean more training for AI tools, and therefore ostensibly better models. Given this requirement for large datasets, it should be of no surprise that the "big players" in this field are those whose business models revolve around the collection of vast amounts of personal data, such as Google, Amazon, Twitter and Facebook.

One such example of the commercialisation of a global userbase is the automated image recognition tool *Google Cloud Vision*, which allows users to "Detect and classify multiple objects including the location of each object within the image" (Google, 2017). Clearly, it would be impossible to "hand code" an algorithm that could identify all possible objects that may appear in an image, so this is a situation where ML models may be ideally employed. However, the "training" of this model requires access to images where objects have *already* been identified. And I am almost certain that you, as an internet user, have been an unknowing part of this training process.

Figure 4.2 shows an example of a *Google reCAPTCHA Enterprise* challenge where, in order to prove that they are not an automated system or "robot", a user must perform identifications on an image. Sometimes this challenge consists of multiple images where one might commonly be asked to identify cars, crosswalks or bicycles. At other times, it is a single image where one might be asked to select

Figure 4.2 An example of a Google reCAPTCHA Enterprise challenge (Image from Google Cloud reCAPTCHA Enterprise documentation and used here under the creative commons attribution 4.0 license).

squares that contain traffic lights or motorbikes. However, this is not solely a tool to differentiate between automatons and organisms; each response to a reCAPTCHA challenge provides identifying information about the objects within images that could not be collected from non-human sources. These images alongside the identifying information can then be used to train the Google Cloud Vision model to provide more accurate responses, improving its commercial value in a way that clearly demonstrates the commodification of an active userbase, again emphasising a recent computer adage that "if using the product is free, then *you* are the product". Resources for teaching about the technical, social and ethical issues around "Big Data" can be found on the *Teach Computing* (2021) website.

Although these forms of ML training can never produce "perfect" models capable of accurate responses all of the time, some of the results are very impressive. The ability of the "botanist in your pocket" app *PictureThis* (Glority, 2018) to correctly identify species of plants from leaves in a given picture, for example, is no mean feat. Whilst we may not be able to penetrate the "hidden layer" between an algorithm's inputs and outputs, we can consider the accuracy and "usefulness" of outputs in relation to inputs. The *Google Neural Machine Translation* system (Wu et al., 2016), for example, created its own language – dubbed interlingua – to translate between human languages, providing a shortcut when introduced to a new language in which it had not yet been "trained". Although this interlingua is not human readable, it is clearly very useful! The language spontaneously developed by Facebook's *Negotiation AI* was perhaps less useful, and the model was shut down when communication outputs deviated from natural speech to the point that developers could not comprehend what was being negotiated between bots (AI Business, 2017), presenting a situation where the connections between input and output were too tenuous to evaluate or provide assurances of safety.

Advertising and content recommendation on social media is an example of clear input-output correlation, as models are trained to predict when people who like/follow/interact with A are more likely to engage with an advert/post/video about B. The rapidly responsive nature of these ever-training algorithms can be seen both in my Spotify recommendations and A. H. Reaume's Instagram adverts. However, these models are extremely susceptible to abuse. The report *Violent Right-Wing Extremism and Terrorism* (Ibsen et al., 2020) outlines the ways that music and violent sports (particularly mixed-martial arts) are utilised as recruitment tools by white supremacist groups. Not only are these connections made at physical sporting and music events but also through digital channels (Zidan, 2018). When members of far-right groups post and interact with hate speech alongside music and sports content, algorithms begin to recommend radicalising videos to fans of particular sports and styles of music, providing digital gateways to extremist content and an audience for recruitment propaganda. Similar patterns of exploiting content recommendation algorithms as a recruitment tool are seen on YouTube for Islamic extremists (Al-Rawi, 2017) and misogynistic "incel" rhetoric (Papadamou et al., 2021), amongst others.

The cumulative effects of algorithmic recommendations lead to "filter bubbles", where sub-groups within an online platform become insulated within their own community (Ananthaswamy, 2011), leading to a narrowed exposure to divergent viewpoints and "confirmation bias" of often controversial or minority viewpoints (Meppelink et al., 2019; Pearson and Knobloch-Westerwick, 2019). Sumpter (2018) recognises not only the prevalence of algorithmically recommended content but also the impact that the resulting filter bubbles can have on meaningful events, such as the 2016 US Presidential Election – arguments that seem to have an even greater gravitas since the widespread dis/mis-information spread around the global coronavirus pandemic.

Sumpter's arguments highlight the role of algorithmic content recommendation in limiting peoples' media consumption to create "filter bubbles", but also emphasises the seemingly hoc way that such imperturbable algorithms ascribe primacy to one post or link over another, leading to an unpredictable (and therefore difficult to manage) relationship between contentious content and user experiences. Indeed, it is accusations of Facebook's role in spreading dis/mis-information, (alongside arguments around exploiting hate speech for profit, and its lack of action despite internal reports outlining its subsidiary Instagram's negative impact on teenage girls' mental health) that led to a recent and extensive rebranding exercise including a name change (Culliford and Dang, 2021).

Issues of content and advertising recommendation further emphasise the need for ethical consideration of algorithms and are driven by digital behaviours. However, given that ML tools are frequently trained on data from the "real world", the resulting outputs can contain existing social biases, often leading to negative social outcomes for already marginalised groups.

The "digital default" and reproduction of bias

In September 2020, under the title of "Trying a horrible experiment...", a Twitter user posted two images, both of which contained photographs of former US president Barack Obama and then Senate Majority Leader Mitch McConnell, but with these portraits arranged in a different order. Both images were too large to be displayed in full on a Twitter feed, so a cropped preview was algorithmically generated by Twitter. Reflecting the results of similar experiments, both previews were cropped to show the white face of Mitch McConnell instead of the arguably far more recognisable black face of Barack Obama. The resulting outcry was rightly reported across many major news outlets (see for example, BBC, 2020; Dasgupta, 2020; Dave, 2021; Knowles, 2020; Metz, 2020), and as a result of Twitter's internal investigations, the algorithm was removed – users are now expected to crop photograph previews themselves (Chowdhury, 2021).

Preferential selection in Twitter preview images is only one of many instances of algorithmic racism when systems are visually presented with non-white skin. Videoconferencing software Zoom is often unable to differentiate black faces from the background, erasing them entirely when a virtual background is used (Dickey, 2020). Google's early image recognition algorithms labelled two black people as "gorillas" (Grush, 2015). Never to be accused of learning from the mistakes of others, six years later Facebook apologised after their algorithm labelled a video of black men as "primates" (Mac, 2021). The fact that across many facial recognition algorithms, lighter-skinned men can be identified with just a 0.7% error rate whereas darker-skinned women have error rates of up to 34.4% speaks for itself, and speaks volumes about training data for and application of visual AI algorithms (Buolamwini and Gebru, 2018). Concerns about how facial recognition technology use in law enforcement might have a disproportionate negative impact on darker-skinned populations prompted technology company IBM to withdraw public access to their identification tools entirely (IBM, 2020).

Further examples of racism in algorithms are easy to find. Whether it's racially biased predictive policing models (Lum and Isaac, 2016), a visa and immigration algorithm described as "speedy boarding for white people" (Joint Council for the Welfare of Immigrants (JWCI), 2020), AI models for the medical diagnosis that are only effective for light skin (Adamson and Smith, 2018) or large racial disparities in algorithmic mortgage lending recommendation rates in the US (Lee and Floridi, 2020; Martinez and Kirchner, 2021), ML models trained on "real-world" data consistently offer negatively skewed outcomes for people of colour.

Even seemingly frivolous algorithms can be radicalised by real-world data. One project you might pursue with your students is building a "chatbot" – an algorithm that asks questions and responds to answers in ways that mimic natural communication (lessons on building a chatbot may be found on TeachWithICT, 2020). In March 2016, Microsoft launched Tay, an AI chatbot that both responds to typed comments from the public and "learning" from these interactions (i.e., further training the ML model). The experiment was shut down after less than 24 hours when Tay's responses began to include racist, sexist and anti-Semitic language. Wolf et al. (2017, p. 1) use Tay to exemplify the fact that if a piece of ML software "…interacts directly with people or indirectly via social media, the developer has additional ethical responsibilities beyond those of standard software". However, this appeal for increased ethical care has not necessarily been widely adopted. Abid et al. (2021) found that GPT-3, an advanced and widely used predictive contextual language model, would associate sentences containing the word "Muslim" with ideas of shooting, bombs, murder and violence over 60% of the time, and "Jewish" was mapped to "money" in 5% of test cases. Gershgorn (2021) reports that GPT-3 was trained on 570 Gb of plain text documents, a dataset "the size of 438,461.5 copies of Moby Dick". Yet the actual content of this dataset is undocumented, leading to a lack of accountability and limited evidence of the required ethical care. In fact, when considering issues with GPT-3, the *MIT Technology Review* found that "undocumented training data perpetuates harm without recourse" (Hao, 2020).

It is not just racial or religious bias that is clearly evident in ML algorithms. As the report *Disability, Bias, and AI* (Whittaker et al., 2019) produced by teams at

New York University and Microsoft shows, disability serves as another form of algorithmic marginalisation. AI-imposed barriers in accessing technology are clearly problematic (such as speech patterns unfamiliar to recognitions models limiting the use of digital personal assistants like Amazon's Alexa or Microsoft's Cortana), but worrying practices such as algorithmic analysis of video job interviews can lead to disability discrimination (Moss, 2020).

Much of the powerful thinking seen in Whittaker et al. (2019) can be applied to consider biases in AI more widely. Where they recognise that those "who have borne discrimination in the past are most at risk of harm from biased and exclusionary AI in the present" (p. 8), they carefully explain how these effects are compounded:

> ...when these discriminatory logics are reproduced and amplified by AI systems, they are likely to be read as authoritative, the product of sophisticated technology. Beyond biased data, additional risks are presented by the significant power asymmetries between those with the resources to design and deploy AI systems, and those who are classified, ranked, and assessed by these systems.
>
> (Whittaker et al., 2019, p. 9)

Such ideas go some way to show why simply adjusting training data to account for existing biases may not be enough, and that developers need to recognise their privileged position whilst attending to the ways that their products may impact upon "target" populations.

Thinking from disability studies also prompts us to consider ideas of "normal" as "normal and abnormal... are often used to distinguish between people with and without disabilities." (Linton, 1998, p. 22). In the intersectional context of AI, we might ask what norms are being constructed within ML models? I want to develop this idea to consider concepts of "digital default" person. The examples already presented in this chapter suggest that to an AI trained on real-world data, the digital default is light-skinned, not Muslim or Jewish, and has no disability. There is also ample evidence that the digital default person is male.

In *Invisible Women*, Criado Perez (2019) demonstrates situation after situation where society assumes a masculine default to the detriment of women, across areas ranging from medical research and employment to architecture and town planning. Such biases are also visible in ML algorithms, including Amazon's AI recruitment selection tool that "taught itself that male candidates were preferable" (Dastin, 2018), the Apple credit card whose automated systems offer up to 20 times the amount of credit to husbands in comparison to their wives (Knight, 2019), Google's Translate algorithm defaulting to masculine pronouns (Zou and Schiebinger, 2018) and YouTube's automated captions struggling to recognise more feminine voices (Tatman, 2017).[1]

OpenAI's ChatGPT (OpenAI, 2023) has become increasingly visible in recent months, providing a natural-language interface to a sophisticated text-based ML model. Despite its meteoric rise in popularity, it is also prone to replication of bias, such as assuming all doctors are male and all nurses are female (Wertheim, 2023). Given the ever-increasing use of ChatGPT in education, be that students generating AI essay drafts or educators using AI-driven learning design, an awareness of these biases is essential.[2]

I have experienced the default male bias in my own research (Dunk and de Freitas, 2019), where I used Amazon's Rekognition tool (Amazon Web Services (AWS), 2016) to analyse photographs of pupils and student teachers in high-school science classrooms. Here, even groups that were 70% female, the algorithm identified just 40% of visible faces as being female, suggesting that any face it cannot identify defaults to male assumptions. This male default is so prevalent in AI systems that it has led to some authors visualising "a future in which artificial intelligence is the ultimate expression of masculinity" (Hayasaki, 2017).

It would be easy to become disheartened with what I have presented thus far in this chapter, seeing the inevitable rise of "Big Data" and the role of algorithms in not just reproducing but exacerbating inequalities. However, I contend that a prerequisite for being able to challenge these issues is first to be aware of this aspect of modern software development. Then we, like others, can work together to raise awareness and challenge these implicit and problematic biases. In the final part of this chapter, I hope to provide some inspiration for what we, as practitioners and educators, might be able to do about it.

Our role as practitioners and educators

As Figure 4.1 suggests, it is impossible to untangle and "police" the hidden layers that ML algorithms build for themselves, so any call to action must be in terms of inputs, outputs and wider social contexts. Dixon-Román (2016, p. 483) argues that as creators and users of algorithms, we must attend to these "relational and connected sociopolitical relations" in our work. Given that implicit algorithmic biases arise as a result of existing biases within society, the most desirable solution is obviously to work to eradicate these biases in general. However, in the absence of profound global progress, we can focus more immediately on building equity and indeed anti-bias into our own digital work and lives, including our teaching practices.

What is obvious is that a better understanding of marginalised groups and how AI impacts them specifically is important. As previously mentioned, an initial first area of focus might be the training data used to build the ML model in the first place. It might be easy to suggest the use of positive discrimination to train an ML model to overcome existing biases as a proactive form of anti-racist/anti-ableist/anti-sexist/etc action, but that in itself is contingent upon being able to identify such biases in the first place. Additional issues arise when developers use "off-the-shelf" ML models where training data are not available. In this case, outputs must be monitored for bias, and once identified, models can be retrained or secondary systems can be constructed to filter or modify outputs before being shared. Such evaluation and feedback mechanisms should be part of the design process, and teaching materials that can support in approaching these issues can be found on the AIinSchools website (Clarke, 2022) or the Apps for Good courses on ML (Apps for Good, 2022).

Part of the argument for increasing diversity in computing classrooms and the workforce is that it will allow for easier identification of bias in the inputs and outputs of ML models. Groups like AllStarCode (2021), Women Who Code (2021), Code2040 (2021), Black Girls Code (2021) and Coding Black Females (2021) alongside projects like Inclusive Coding for Disabled Developers (BCU, 2021) are working in this area, but we must be mindful of the importance of diversity in the workforce as an essential component in challenging bias. As Bloodhart et al. (2020) found in considering gender biases in STEM fields:

> …women's performance and accomplishments in STEM need to be publicly recognized to address misperceptions that women are less capable, skilled, or have less expertise.
>
> (Bloodhart et al., 2020, p. 10)

Similar arguments can easily be made for celebrating and highlighting the successes of other marginalised groups as one route towards improving the experiences of those groups within computing fields. However, approaches that simply acknowledge the successes of a diverse workforce are by no means sufficient. Putting the "work" of addressing bias on marginalised is an unfair burden. Instead, we must find ways to engage *all* groups in facing up to their own prejudices, biases and privilege.

One such approach to building more inclusive computing classrooms is to employ culturally relevant pedagogy (Ladson-Billings, 1995) in our teaching and training, and much has been written on the use of such approaches in the context of computer science (e.g., Coleman, 2021, p. 34; Madkins et al., 2019; Mejias et al., 2018; Rorrer et al., 2018). Such an approach draws on learners' cultural knowledge and experiences to inform the curriculum whilst exploring issues of social justice and bias. Some simple, practical ways to make computing lessons accessible to those with additional needs can be found on Hello World (Elliott, 2020), and examples of how technology provides personal solutions to learning issues can offer important teaching points, such as the "one-handed" bass guitar (BBC Bitesize, 2020). Fiesler (2021) provides a comprehensive list of resources of openly available ethical computing syllabi that you might like to incorporate into your own work.

Although ethical issues are not mentioned in the Computing National Curriculum for England and Wales, and only rarely appear in examined content, I hope this chapter has served to highlight the importance of addressing such issues in the classroom for future programmers, for digitally informed citizens and for society more generally. The responsibility for recognising, acknowledging and challenging the biases within computing both as a field of study and within the outputs of those working in the field (algorithms, software, etc) lies with all of us. Seize that challenge.

Final reflection points

1. Consider a long-term curriculum plan for computing in your establishment. Can you identify areas where you teach issues around ethical computing? What opportunities are there to increase the profile of these issues throughout all lessons on the plan?

2. What are you doing to celebrate the successes of marginalised groups in your teaching? Are there opportunities in your long-term plan to better raise the profile of women, people of colour or those with disabilities working in computing fields? This shouldn't be taken as a token nod towards political correctness but as an imperative step towards addressing inequality in the subject area and products of the subject area more broadly.

3. Look at the composition of your computing classes. Are there any definite patterns of bias? How might you work to mitigate these biases, either in your recruitment practices or ongoing teaching of the subject?

Notes

1 It is worth acknowledging the geographic differences in speech recognition. A quick search using the terms "Scottish Alexa" produces many videos of people struggling to be understood by their digital assistant, suggesting that our digital default is also unlikely to have a strong regional accent.
2 Ferrara (2023) provides a detailed analysis of the types of bias found in ChatGPT, and how these biases can be identified, quantified, and mitigated in ML models.

References

Abid, A., Farooqi, M., Zou, J., 2021. Persistent anti-Muslim bias in large language models. arXiv:2101.05783 https://doi.org/10.48550/arXiv.2101.05783.

Adamson, A.S., Smith, A., 2018. Machine learning and health care disparities in dermatology. *JAMA Dermatology* 154, 1247–1248. https://doi.org/10.1001/jamadermatol.2018.2348

AI Business, 2017. Facebook shuts down AI negotiating bot after it invented its own language [WWW Document]. AI Business. URL https://www.aibusiness.com/document.asp?doc_id=760394 (accessed 10.30.21).

AllStarCode, 2021. Letter from our founder. URL https://www.allstarcode.org/a-letter-from-our-founder/ (accessed 11.1.21).

Al-Rawi, A., 2017. Introduction, in: Al-Rawi, A. (Ed.), *Islam on YouTube: Online Debates, Protests, and Extremism.* Palgrave Macmillan, London, pp. 1–6. https://doi.org/10.1057/978-1-137-39826-0_1

Ananthaswamy, A., 2011. Age of the splinternet. *New Scientist* 211, 42–45. https://doi.org/10.1016/S0262-4079(11)61710-7

Apps for Good, 2022. Machine learning standard [WWW Document]. Apps for Good. URL https://www.appsforgood.org/courses/machine-learning (accessed 2.24.22).

Amazon Web Services (AWS), 2016. Rekognition [WWW Document]. URL https://aws.amazon.com/rekognition/ (accessed 4.18.18).

BBC, 2020. Twitter investigates racial bias in image previews. BBC News.

BBC Bitesize, 2020. Computer science KS3 & 4/GCSE: problem solved – one handed bass [WWW Document]. BBC Class Clips Video. URL https://www.bbc.co.uk/teach/class-clips-video/problem-solved-one-handed-bass/z6ft382 (accessed 2.24.22).

BCU, 2021. Inclusive coding for disabled developers [WWW Document]. URL https://www.bcu.ac.uk/computing/research/digital-media-technology/research-projects/coding-for-disabled-developers (accessed 11.1.21).

Black Girls Code, 2021. About us [WWW Document]. Black girls code—women of color in technology. URL https://www.blackgirlscode.com (accessed 11.1.21).

Bloodhart, B., Balgopal, M.M., Casper, A.M.A., McMeeking, L.B.S., Fischer, E.V., 2020. Outperforming yet undervalued: undergraduate women in STEM. *PLOS ONE* 15, e0234685. https://doi.org/10.1371/journal.pone.0234685

Buolamwini, J., Gebru, T., 2018. Gender Shades: Intersectional Accuracy Disparities in Commercial Gender Classification, in: Proceedings of the 1st Conference on Fairness, Accountability and Transparency. Presented at the Conference on Fairness, Accountability and Transparency, PMLR, pp. 77–91.

Chowdhury, R., 2021. Sharing learnings about our image cropping algorithm [WWW Document]. URL https://blog.twitter.com/engineering/en_us/topics/insights/2021/sharing-learnings-about-our-image-cropping-algorithm (accessed 10.30.21).

Clarke, B., 2022. AIinSchools – utopia vs dystopia [WWW Document]. URL http://aiinschools.com/ (accessed 2.24.22).

Code2040, 2021. Mission [WWW Document]. URL http://www.code2040.org/mission (accessed 11.1.21).

Coding Black Females, 2021. About [WWW Document]. URL https://codingblackfemales.com/about (accessed 11.1.21).

Coleman, G. (Ed.), 2021. *The Big Book of Computing Pedagogy, (Hello World)*. Raspberry Pi Press, Cambridge.

Criado Perez, C., 2019. Invisible women: exposing data bias in a world designed for men. Chatto & Windus.

Culliford, E., Dang, S., 2021. Facebook changes name to Meta as it refocuses on virtual reality. Reuters.

Dasgupta, B., 2020. Racist machines? Twitter's photo preview problem reignites AI bias concern [WWW Document]. Hindustan Times. URL https://www.hindustantimes.com/india-news/racist-machines-twitter-s-photo-preview-problem-reignites-ai-bias-concern/story-kYWMvpx1WaV77NiVOx8KDP.html (accessed 10.30.21).

Dastin, J., 2018. Amazon scraps secret AI recruiting tool that showed bias against women. Reuters.

Dave, P., 2021. Twitter finds its AI tends to crop out Black people, men from photos. Reuters.

Dickey, M.R., 2020. Twitter and Zoom's algorithmic bias issues. TechCrunch.

Dixon-Román, E., 2016. Algo-Ritmo: more-than-human performative acts and the racializing assemblages of algorithmic architectures. *Cultural Studies ↔ Critical Methodologies* 16, 482–490. https://doi.org/10.1177/1532708616655769

Doffman, Z., 2019. Hong Kong exposes both sides of China's relentless facial recognition machine. Forbes.

Dunk, R.A., de Freitas, E., 2019. Ghosts, Zombies and Other Spooky Creatures: New Methods for Visualizing Agency and 'Presence' in Classrooms. Presented at the AERA Annual Meeting, Toronto.

Elliott, Catherine., 2020. The inclusive computing classroom [WWW Document]. Hello World. URL https://helloworld.raspberrypi.org/articles/HW11-the-inclusive-computing-classroom (accessed 2.24.22).

Ferrara, E., 2016. Should ChatGPT be biased? Challenges and risks of bias in large language models. *arXiv 2304.03738*. https://doi.org/10.48550/arXiv.2304.03738

Fiesler, C., 2021. Tech ethics curricula: a collection of syllabi. Medium. URL https://cfiesler.medium.com/tech-ethics-curricula-a-collection-of-syllabi-3eedfb76be18 (accessed 2.24.22).

Gautam, T., 2020. Image classification in python with keras. Analytics Vidhya. URL https://www.analyticsvidhya.com/blog/2020/10/create-image-classification-model-python-keras/ (accessed 2.16.21).

Gershgorn, D., 2021. GPT-3 contains disturbing bias against Muslims. OneZero. URL https://onezero.medium.com/for-some-reason-im-covered-in-blood-gpt-3-contains-disturbing-bias-against-muslims-693d275552bf (accessed 10.31.21).

Glority, 2018. PictureThis - plant identifier [WWW Document]. URL https://www.picturethisai.com/ (accessed 10.30.21).

Google, 2016. Quick, Draw! [WWW Document]. URL https://quickdraw.withgoogle.com/ (accessed 7.26.21).

Google, 2017. Cloud vision API [WWW Document]. URL https://cloud.google.com/vision/ (accessed 4.18.18).

Grush, L., 2015. Google engineer apologizes after photos app tags two black people as gorillas. The Verge.

Hao, K., 2020. We read the paper that forced Timnit Gebru out of Google. Here's what it says. MIT Technology Review.

Hayasaki, E., 2017. Is AI sexist? Foreign Policy.

IBM, 2020. IBM CEO's letter to congress on racial justice reform [WWW Document]. URL https://www.ibm.com/blogs/policy/facial-recognition-sunset-racial-justice-reforms/ (accessed 10.31.21).

Ibsen, D., Pham, L., Schindler, H.-J., Ritzmann, A., Rekawek, K., Fisher-Birch, J., Macori, M., 2020. *Violent Right-Wing Extremism and Terrorism – Transnational Connectivity, Definitions, Incidents, Structures and Countermeasures.* Counter Extremism Project, Berlin.

Joint Council for the Welfare of Immigrants (JWCI), 2020. We won! Home office to stop using racist visa algorithm [WWW Document]. Joint Council for the Welfare of Immigrants. URL https://www.jcwi.org.uk/news/we-won-home-office-to-stop-using-racist-visa-algorithm (accessed 10.30.21).

Knight, W., 2019. The Apple card didn't "see" gender—and that's the problem. Wired.

Knowles, T., 2020. Twitter admits its image algorithm was racist.

Ladson-Billings, G., 1995. Toward a theory of culturally relevant pedagogy. *American Educational Research Journal* 32, 465–491. https://doi.org/10.3102/00028312032003465

Lee, M.S.A., Floridi, L., 2020. Algorithmic fairness in mortgage lending: from absolute conditions to relational trade-offs. *Minds and Machines* 31, 165–191.

Linton, S., 1998. *Claiming Disability: Knowledge and Identity (Cultural Front): 14.* NYU Press, New York.

Lum, K., Isaac, W., 2016. To predict and serve? *Significance* 13, 14–19. https://doi.org/10.1111/j.1740-9713.2016.00960.x

Mac, R., 2021. Facebook apologizes after A.I. puts 'primates' label on video of black men. The New York Times.

Madkins, T.C., Martin, A., Ryoo, J., Scott, K.A., Goode, J., Scott, A., McAlear, F., 2019. Culturally Relevant Computer Science Pedagogy: From Theory to Practice, in: 2019 Research on Equity and Sustained Participation in Engineering, Computing, and Technology (RESPECT). Presented at the 2019 Research on Equity and Sustained Participation in Engineering, Computing, and Technology (RESPECT), pp. 1–4. https://doi.org/10.1109/RESPECT46404.2019.8985773

Martinez, E., Kirchner, L., 2021. The secret bias hidden in mortgage-approval algorithms. The Markup.

Mejias, M., Jean-Pierre, K., Burge, L., Washington, G., 2018. Culturally Relevant CS Pedagogy – Theory amp; Practice, in: 2018 Research on Equity and Sustained Participation in Engineering, Computing, and Technology (RESPECT). Presented at the 2018 Research on Equity and Sustained Participation in Engineering, Computing, and Technology (RESPECT), pp. 1–5. https://doi.org/10.1109/RESPECT.2018.8491699

Meppelink, C.S., Smit, E.G., Fransen, M.L., Diviani, N., 2019. "I was right about vaccination": confirmation bias and health literacy in online health information seeking. *Journal of Health Communication* 24, 129–140. https://doi.org/10.1080/10810730.2019.1583701

Metz, R., 2020. Twitter looking into racial bias in tweet image previews [WWW Document]. CNN. URL https://www.cnn.com/2020/09/21/tech/twitter-racial-bias-preview/index.html (accessed 10.30.21).

Mohri, M., Rostamizadeh, A., Talwalkar, A., 2018. *Foundations of Machine Learning*, second edition. MIT Press. Cambridge, Massachusetts.

Moss, H., 2020. Screened out onscreen: disability discrimination, hiring bias, and artificial intelligence. *Denver Law Review* 98, 775–806.

OpenAI, 2023. Introducing ChatGPT [WWW Document]. URL https://openai.com/blog/chatgpt (accessed 6.22.23).

Papadamou, K., Zannettou, S., Blackburn, J., De Cristofaro, E., Stringhini, G., Sirivianos, M., 2021. "How over is it?" Understanding the Incel Community on YouTube. *Proceedings of the ACM on Human-Computer Interaction* 5, 1–25. https://doi.org/10.1145/3479556

Pearson, G.D.H., Knobloch-Westerwick, S., 2019. Is the confirmation bias bubble larger online? Pre-election confirmation bias in selective exposure to online versus print political information. *Mass Communication and Society* 22, 466–486. https://doi.org/10.1080/15205436.2019.1599956

Purshouse, J., Campbell, L., 2019. Privacy, crime control and police use of automated facial recognition technology. *Criminal Law Review* 2019, 188–204.

Reaume, A.H., 2020. @a_h_reaume [WWW Document]. URL https://twitter.com/a_h_reaume/status/1336833981422084098 (accessed 10.30.21).

Rorrer, A.S., Allen, J., Zuo, H., 2018. A National Study of Undergraduate Research Experiences in Computing: Implications for Culturally Relevant Pedagogy, in: Proceedings of the 49th ACM Technical Symposium on Computer Science Education, SIGCSE '18. Association for Computing Machinery, New York, NY, pp. 604–609. https://doi.org/10.1145/3159450.3159510

Sumpter, D., 2018. *Outnumbered: From Facebook and Google to Fake News and Filterbubbles – The Algorithms That Control Our Lives*. Bloomsbury Sigma, London.

Tatman, R., 2017. Gender and Dialect Bias in YouTube's Automatic Captions, in: Proceedings of the First ACL Workshop on Ethics in Natural Language Processing. pp. 53–59.

Teach Computing, 2021. Lesson 3 big data [WWW Document]. Teach Computing. URL https://teachcomputing.org (accessed 2.24.22).

TeachWithICT, 2020. Python chatbot tutorial [WWW Document]. teachwithict.com. URL https://www.teachwithict.com/chatbot.html (accessed 1.29.22).

Waddell, K., 2019. Utilities are turning to AI to predict coming disasters [WWW Document]. Axios. URL https://www.axios.com/predictive-maintenance-utility-power-gas-2db85586-cf32-4d81-baaf-f4720bbb2d73.html (accessed 7.26.21).

Wertheim, S., 2023. ChatGPT insists that doctors are male and nurses female [WWW Document]. URL https://www.worthwhileconsulting.com/read-watch-listen/chatgpt-insists-that-doctors-are-male-and-nurses-female (accessed 6.22.23).

Whittaker, M., Alper, M., College, O., Kaziunas, L., Morris, M.R., 2019. Disability, bias, and AI. New York University (NYU).

Wolf, M., Miller, K., Grodzinsky, F., 2017. Why we should have seen that coming: comments on Microsoft's Tay "experiment," and wider implications. *The ORBIT Journal* 1, 1–12. https://doi.org/10.29297/orbit.v1i2.49

Wolkenstein, A., 2018. What has the Trolley Dilemma ever done for us (and what will it do in the future)? On some recent debates about the ethics of self-driving cars. *Ethics and Information Technology* 20, 163–173. https://doi.org/10.1007/s10676-018-9456-6

Women Who Code, 2021. About us [WWW Document]. URL https://www.womenwhocode.com/about (accessed 11.1.21).

Wu, Y., Schuster, M., Chen, Z., Le, Q.V., Norouzi, M., Macherey, W., Krikun, M., Cao, Y., Gao, Q., Macherey, K., Klingner, J., Shah, A., Johnson, M., Liu, X., Kaiser, Ł., Gouws, S., Kato, Y., Kudo, T., Kazawa, H., Stevens, K., Kurian, G., Patil, N., Wang, W., Young, C., Smith, J., Riesa, J., Rudnick, A., Vinyals, O., Corrado, G., Hughes, M., Dean, J., 2016. Google's neural machine translation system: bridging the gap between human and machine translation. arXiv:1609.08144. https://doi.org/10.48550/arXiv.1609.08144

Yala, A., Lehman, C., Schuster, T., Portnoi, T., Barzilay, R., 2019. A deep learning mammography-based model for improved breast cancer risk prediction. *Radiology* 292, 60–66. https://doi.org/10.1148/radiol.2019182716

Zidan, K., 2018. Fascist fight clubs: how white nationalists use MMA as a recruiting tool. The Guardian.

Zou, J., Schiebinger, L., 2018. AI can be sexist and racist—It's time to make it fair. *Nature* 559, 324–326. https://doi.org/10.1038/d41586-018-05707-8

Using data to ensure an engaging and inclusive computing curriculum

Matthew Thorpe

Introduction

Data are increasingly being used for analysis across all parts of society. Our digital footprints are analysed so we can be provided with bespoke services and recommendations when we use social media, streaming services and do our online shopping. Data-informed products and services have created many new opportunities and also challenges societally as we spend more of our time online. Education is in no way different from this wider reality; by increasingly capturing educational data, this provides both opportunities and challenges that teachers must be aware of. This chapter will explore specific data-related examples and applications and enable you to consider how educators can harness the educational benefits of data whilst being mindful of associated ethical dilemmas.

A short history of data in UK schools

Data is a term we hear regularly in educational settings today, but before we progress let's consider what this word actually means. Data can be defined simply as the recording of information and statistics for analysis. It will be useful for you to remember this broader classification of data throughout this chapter, as this will allow you to think more flexibly about where you record data in your practice, and the insight this offers you about your learners.

Due to technological advances, many of us today have a narrower conception of data as being digitally stored on computer networks, accessible to teachers via student record systems in schools, for example, summative assessment data. However, the term has a longer history that predates digital technologies considerably. Since the 19th century, it has been compulsory for schools in England and Wales to keep records of attendance data (Stephens, 1998). Similarly, schools have kept records of assessment data for well over a century. The 1980s marked an acceleration in the recording of data as student attainment became much more central in educational policy and inspection frameworks. This increased accountability for schools has led to the more detailed and varied recording of student data within schools in recent decades.

In recent years, schools have recorded increasing amounts of data digitally. Ofsted's report "Using data, improving schools" (Ofsted, 2008) represents a key development in policy, designed to support schools in harnessing the potential of digital data for school improvement, and to help senior leaders and teachers understand its limitations. Most recently, data have experienced increased attention in schools due to the new General Data Protection Regulation (GDPR) legislation, which came into UK law in 2018. This has led to the creation of a GDPR toolkit (Department for Education, 2018) designed to support schools in achieving compliance when handling staff and student data.

As the volume and granularity of student data have increased, this has created new dilemmas for the education sector in terms of how data can be used to promote rather than hinder inclusive practices. For example, does exposure to assessment data – visible to teachers, students and parents – act as actionable intelligence to support underperforming students and promote inclusion? Or can such data reinforce underperformance and become a self-fulfilling prophecy for students? Similarly, do increasing amounts of data outlining learning needs for SEND or pupil premium children lead to educators being able to offer more differentiated support? Or can it lead to unfair labelling and assumptions about students? These types of questions require careful consideration within school settings, and increased data literacy from teachers and senior leaders can help ensure these complex data dilemmas are dealt with in an appropriate and ethical manner to maximise inclusivity.

Much of what we have discussed historically thus far relates to the use of data at a policy, national and school level, for example, the recording of attendance and summative assessment data. However, over recent years, there has been an explosion in the number of educational applications and technologies teachers can now use with their classes. These tools create a significant amount of data that can be analysed by teachers and learners to promote effective learning. For example, the Eedi platform, as seen here: https://eedi.com/, is used by the National Centre for Computing to provide computing teachers with diagnostic and summative assessment data about their students. This enables teachers to better understand learner progress and devise support strategies for their classes.

Why are data important for teachers and learners?

More student data are being generated than ever before thanks to advances in hardware, software and cloud computing (Nesta, 2016). Many of the educational technologies we will be discussing in this chapter automatically capture data as students interact with them; this data can then be analysed by teachers for a variety of educational reasons. This auto-generated data are useful in that it does not create further administrative burden for teachers; rather, it is a case of teachers having sufficient data literacy skills to analyse the data in meaningful ways where appropriate. Increasingly varied data offer benefits beyond just improving attainment (Nesta, 2016); it allows teachers to get to know students as individuals far better. This more nuanced understanding allows teachers to detect whether students are struggling more quickly and allows them to make more evidence-based decisions in their planning and teaching.

Data enable lessons to be tailored to the individual needs of their students (Yin, 2019). By seeing students as individuals, this encourages teachers to embed more effective formative assessment and differentiation strategies in their practice. Data also allows teachers to consider aspects beyond just the academic progress of students. It can be analysed to offer insight into students' affective states and wellbeing (Ahern, 2018). If analysis does raise a concern, pastoral care or specialist support teams with expertise in mental health and counselling can intervene in a timelier manner with students.

> **Activity**
>
> We will explore some specific technologies and examples of how data can be used by teachers to support students more effectively further in this chapter. However, now is a good time to start thinking about how you could utilise data in your own practice to support your students? What technologies or applications might you use? What data would be useful to capture? What benefits would this allow you and your students?

What data do teachers come into contact with?

It is important to have a broad awareness of the varied types of data teachers produce, consume and engage with frequently. As we have discussed, teachers are steered by more formal quantitative data such as attendance data, assessment marks and target grades; however, there are many more sources of data beyond this.

Teachers produce qualitative data in the shape of progress reports, tutorial notes and observational data in class files. Most of this data is now stored digitally for easy retrieval and analysis. Many schools also allow students to narrate their own progress in the form of self-reports. This type of data is equally insightful in providing context to support students. There is much more significant variance in how teachers use data captured by specific applications at the subject or class level. Teachers now use a whole host of educational technologies to increase student engagement during lessons; for example, a teacher might do an online quiz at the end of each class. Alongside this, students might be required to access and engage with resources remotely as homework to supplement face-to-face teaching.

These tools generate data that can be exported and visualised in various formats (spreadsheets, CSV, PDF, graph charts and line charts). Many teachers do not access and analyse this data, quite often due to a lack of awareness that any data has been captured. This represents a missed opportunity for teachers to more effectively formatively assess student progress and better understand their classes' individual learning needs.

Apps that enable data collection and analysis

Integrated development environments (IDEs) are applications that allow students to develop their programming skills. Traditionally, a computing teacher may access data such as written code, comments and compile errors to assess student progress from an IDE. However, IDEs have been trialled in recent years that generate more varied data insights, allowing a fuller picture of learner progress (Hundhausen et al., 2017). By embedding social media-style spaces into such platforms, this enables students to debug code socially with their peers. This qualitative data are also visible to the teacher, who can view when and how often learners post and reply to each other's questions. By capturing data in this way, the teacher can build a fuller picture of learner progress and provide more immediate and bespoke feedback via the chat space in the IDE itself. Digital badges can also be awarded to students in the social space within the IDE, which recognises and rewards programming skills acquired.

Future work in this space is now looking at how IDEs can be augmented further by collecting data in response to quiz questions to provide more detailed insights for analysis. Physiological data such as eye tracking and heart rate are also being explored, which would not only offer another dimension in measuring engagement and learning during programming tasks but also the physical and mental well-being of students. For example, much varied work has already established a relationship between erratic eye movements and a poor mental state. iSnap represents another example of an application exploring how data can be used to better support students in developing their programming skills. iSnap is an online block-based programming environment that has similar functionality to Scratch. Such platforms represent a great way of transitioning students to text-based programming using a language like Python, for example. iSnap has been developed by researchers at Berkeley University and uses data in new and innovative ways to support learners with their programming. The tool collects data from students as they complete programming tasks, and these data are then used to generate on-demand hints and next steps when students get stuck on a task (Price et al., 2017). Students can request hints and hover over blocks of their code when they are struggling to receive tailored support. iSnap offers an example of how data can be captured and analysed to enable automated feedback and support from within the programming platform itself. It can be found here: https://snap.berkeley.edu/

> **Activity**
>
> Take a look at the "Ten quick tips for teaching programming" paper by Brown and Wilson (2018) in the reference list. This is a good general resource for thinking about effective pedagogy for programming lessons. Within the context of this chapter, though, think about the following based on the two examples we have just discussed:
>
> - How does the use of data to support students programming align with the ten tips offered in the paper? For example, one tip discusses the advantages of peer programming, how does this compare with the example of students discussing their code with one another in an integrated chat space within the IDE?
> - What benefits does an increasing amount of data enable in supporting learners to develop programming skills? For example, can data help to better differentiate programming activities to ensure they are more inclusive?
> - Do the examples discussed raise any concerns from a teacher's perspective? For example, would some teachers have an issue with students receiving automated prompts and hints within the IDE?

Blended and online learning

An increase in online learning platforms represents a shift in pedagogy both online and in the physical classroom. These platforms allow pupils to be formatively assessed via online quizzes, which can often be configured to auto-mark and send feedback to students, enabling teachers to embed assessments without needing to consume too much of their time with marking. Such data can be analysed by the teacher and students themselves as a useful predictor for pupil performance in final exams. This data can of course also be used to alter these predicted trajectories. For example, if data analysis shows some students are on track to perform poorly in a specific topic area, this allows the teacher to revisit the topic in class, but it may also mean pupils need to conduct further independent study or revision around the topic.

Pupils can also complete surveys via online forms. The data allow the teacher to drill into individual responses or export to a spreadsheet for manipulation and analysis of larger data sets. Most online forms automatically create graphical representations of data that can be displayed in bar charts, pie charts, etc.

Online learning platforms represent a shift to cloud storage (OneDrive and Google Drive) and online collaboration. Teachers now have access to learners' work via the web, which they can use to view progress at any time they wish, for example to make comments or track changes to an online Word document in progress, rather than waiting for the student to submit work for review or marking more formally. Online files are also great for project work and peer assessment, enabling pupils to collaborate and comment on each other's shared files remotely.

Video conferencing tools such as Microsoft Teams and Google Meet also enable data analysis. In addition to observing pupils' contributions on the video call itself, teachers can consider chat contributions, whiteboards and polls to assess engagement and learning. Beyond this, applications include increasingly sophisticated data analytics capabilities. For example, individual user data on the following:

- mute/unmute activity.
- when pupils arrive and leave online sessions.
- number of messages posted in chat.
- number of 1:1 calls.
- number of meetings setup.
- last activity date of a user.
- access platform (Windows/Mac/iOS/Android).
- screen sharing time.

By correlating and analysing this type of pupil data during individual classes and across full terms, this can help the teacher develop an intricate picture of pupil engagement, which represents a good predictor for overall outcomes. These types of data have been especially useful for gauging learner progress during remote learning; however, chat functions and collaboration on cloud-based files are now also being used alongside face-to-face teaching in class, meaning the data also represent a good indicator of pupil progress whilst in school also.

Statistical data, typically in the shape of formative and summative assessments, can be easily collated within a digital gradebook on most online learning platforms; this allows for automatic calculation of overall grades and averages over the course of a topic or academic year. Customisation also enables the teacher to switch between percentages, marks or assign custom weightings to different assessments throughout the year. Data from the gradebook are usually easily exportable as a CSV or Excel file for record-keeping or more detailed analysis. Teachers can often be teaching four or five different classes a day, seeing more than 150 pupils, so the ability to configure a custom gradebook can save time that would be taken manually recording assessment data. Use of an integrated gradebook also enables teachers to have a much better overview of pupil progress throughout the year, which allows for the implementation of more effective assessment for learning (AFL) strategies. To ensure teachers can utilise data effectively, schools must ensure appropriate training is available for staff to enable the development of data literacy. Having digital experts available in school to provide support is another model that has proven to be effective, although this practice is more common in higher education settings at present.

Although blended and online learning offers exciting possibilities to better support learners in offering a more differentiated and bespoke education, we must be mindful of the fact that equally it has at times led to inequalities in experience amongst pupils. To access online technologies remotely when not at school, pupils need reliable access to the web and hardware such as laptops or tablets. It is important to remember that due to socio-economic factors, often pupils are in a situation where they may not have a device or must share one with siblings. Pupils may also not have an appropriate study space at home free from distractions. Many of these issues became increasingly prominent during the Covid-19 pandemic, with national school closures and the move to online education. This led to the DfE offering laptop loans and 4G wireless routers for disadvantaged pupils. In the long term, even with the return of face-to-face teaching, there is increasingly an expectation, especially in technology-based subjects that pupils will need to access homework and further support materials online. We must ensure we do not create a "digital divide" (Van Dijk, 2020) in our education system and be responsive to the inequities that exist amongst learners concerning home use of technology. Such inequities are most visible in relation to pupil access to appropriate hardware, software and the web. However, we also must consider the varied amounts of knowledge and support pupils have access to in the family home due to parents/carers having diverse digital literacies themselves (Coleman, 2021). Some parents may have limited digital skills, and in other circumstances, they will also have limited literacy skills or not speak English as a first language. These inequalities will only become heightened as pupils are expected to produce larger amounts of online work remotely. We must therefore be aware of such scenarios and ensure we are inclusive of all pupils.

> **Activity**
>
> Think about any personal experiences you have with teaching or learning online. How can online learning platforms be used effectively both within the classroom and remotely to best support computing pupils? What safeguards can be put in place from a subject perspective to reduce the impact of the digital divide and promote an inclusive curriculum?

Promoting an inclusive and engaging classroom with educational technologies and data

In addition to the usefulness of technology for online learning, there are a whole host of applications that can be used by teachers in the classroom to make learning interactive, fun, and engaging. Such technologies usually require the pupils to interact with questions and discussions via a laptop, tablet or smartphone. Unlike more traditional questioning techniques used by teachers, the benefit of using technologies like these is that all pupils must provide an answer via their device rather than just the individual to whom a question is directed. These technologies help promote inclusion for all, receiving much positive feedback from learners who may want to contribute, but be uncomfortable speaking in front of the rest of the class.

Valuable AFL data are produced by such interactive classroom technologies, this data can be displayed immediately after pupils have inputted their answer, which enables them to self-assess their knowledge and where they might need to further develop. At a class level, the teacher can consider this data in terms of future planning, for example, if the class has performed particularly poorly in one area, a topic might need further coverage. Teachers can also use individual pupil data to differentiate support, ensuring lessons are planned that are accessible and inclusive for all learners.

> **Reflection Points**
>
> Below (Table 5.1), three specific classroom technologies are provided as examples, although there are many more! Whilst looking at these examples, consider how you would embed such technologies into your lesson planning to promote inclusion and engagement in your teaching? Do some research on the web to look at other applications you could also use in your practice?

Exploring the intersections of learning theory and data-informed practice

Although we have largely focused on practical examples of data-informed practice in this chapter, it is worth briefly considering the implications data has for learning at a more theoretical level. As we have seen, data typically captures explicit pupil behaviours and actions, for example, clicks and posts in various software applications, attendance and assessment scores. These behaviours can then be analysed as proxies for pupil engagement, learning and well-being.

Due to this, data-informed pedagogy tends to most obviously encourage behaviourst theories of learning. This theory dates to the early 20th century and was developed alongside an increasingly scientific approach to studying human behaviour. When adopting a behaviourist approach, only observable phenomena are of interest, typically in the form of changes within the subject's behaviour based upon a specific stimulus. An individual's internal sense-making and processing are within the "black box" of consciousness. As a result, these phenomena are not observable and therefore not of interest to behaviourists.

Table 5.1 Interactive classroom technologies that promote assessment for learning (AFL), inclusion and engagement

Category	Application	Discussion
Video quizzing tools	EdPuzzle www.edpuzzle.com	Video hosting platforms are also now capable of generating insightful data. EdPuzzle is a video hosting and quizzing tool via which teachers can upload video content with inbuilt quiz questions for pupils to engage with. EdPuzzle provides data that enables teachers to see if and when a pupil has watched a video, how many times they have watched particular sections of the video, any answers they have submitted and how long they spent engaged with the activity. Individual or class-level data can be analysed by the teacher, and this insight can be used to send bespoke feedback to pupils via the platform. In comparison to non-data-informed uses of educational videos, EdPuzzle allows the teacher to intervene with pupils who are not fully engaging, providing more bespoke feedback and support. It also allows the teacher to reflect on the relevance of the materials they are producing. For example, if one video has particularly low engagement, a teacher may decide to replace it with something more meaningful. In this context, the data can inform curriculum design. Question types include open-ended questions, multiple-choice questions, and the ability for pupils to make notes at certain points in the video. Take a look at some of the video quizzes shared by others on the EdPuzzle website; this will give you a feel for how the tool works.
Quizzing tools	Kahoot www.kahoot.com	Quizzing tools like Kahoot offer a further dimension by adding a gamified and competitive element to learning. This is achieved via live data, which is fed back to the pupils in the form of a leader board, after each question pupils receive a certain amount of points, based upon whether they have given a correct response and the speed of answer. Kahoot and similar tools have proved to be an easy and fun way to embed AFL within lessons regularly. Teachers can also export a more detailed spreadsheet of responses for later analysis, this can enable them to construct specific support strategies for individual pupils, and shape planning for upcoming lessons. Kahoot mainly uses multiple choice question types; however, true or false, typed answers and puzzle questions such as sorting multiple answers into a specific sequence can be added. Make sure to visit the Kahoot website, which contains a large library of pre-created quizzes shared by others that you can edit and use yourself.

(Continued)

Table 5.1 (Continued)

Category	Application	Discussion
Interactive presentation applications	Mentimeter www.mentimeter.com	Interactive presentation applications work similarly to Kahoot but allow for a wider variety of responses. For example, longer free text prompts or questions allow pupils to provide more opinion-based responses rather than a correct/incorrect solution. These question types can be great for prompting further discussion and debate in class. The textual data produced can be analysed by the teacher to assess pupils' rationales and understandings of specific topics. In addition to this data providing insight into pupils' progress academically, it can also provide sentiment analysis. This offers an overview of how pupils are feeling about a particular topic or on a particular day. Word cloud questions are a great way of doing this. The more frequently the class contributes a particular word, the more prominent it becomes in the word cloud. If you are thinking of trying a pupil response system in your practice, Mentimeter and NearPod are popular examples of such tools. Check out the "inspiration" page on the Mentimeter website to see live examples and suggestions for how to use the tool to maximise engagement.

Operant conditioning represents a notable development within the field of behaviourism. Demonstrated by B.F. Skinner's study involving rats, which investigated how behaviour can be influenced using reward or punishment, if a rat pressed one lever, they would receive food; the other lever would result in the rat getting an electric shock. Skinner's work demonstrated how this approach could condition the rats to change their behaviours. So how does the notion of conditioning relate to educational technologies and the data they produce?

Many educational technologies today measure pupils' behaviours as data, using similar principles of conditioning to promote certain types of behaviour. For example, attendance-monitoring systems will often automatically send parents and/or pupils an email or text message if attendance drops below a certain threshold. Equally, the system may send a communication offering praise for learners who have maintained 100% attendance over a term. These communications hope to prompt a certain type of behaviour from pupils based on data analysis.

ClassDojo is a platform widely used in primary schools that allows teachers, pupils and parents to share messages, photos and videos as evidence of the learning that has taken place. The tool also enables teachers to award plus or minus "Dojo points" to pupils' profiles to promote positive pupil behaviour in class. Quizzing tools can also be set to automatically provide feedback to pupils based on their responses to multiple-choice questions, offering praise for a correct answer or further guidance for an incorrect answer. These tools can be configured to use branching to steer pupils to easier or more difficult resources, depending on their responses or behaviours.

Data can additionally gamify learning by adding a competitive element, such as leader boards. Kahoot is a good example of this, where a correct answer and then the speed of response are correlated to give each pupil a certain number of points. After each question, the pupils see a leader board, and on completion of the quiz, the top three pupils are displayed on a podium. By gamifying learning, tools like Kahoot are designed to encourage pupils to be more engaged and focused in their learning.

Although apps like ClassDojo and Kahoot represent specific technologies that many teachers find highly effective in promoting engagement and positive behaviours in the classroom, it is important that we do not become overly reliant on such approaches with pupils. We must ensure that rather than relying on technologies that promote behaviourist notions of learning, these are used in a complementary fashion alongside opportunities for socially mediated, deeper and more meaningful educational experiences.

Computing has a rich history of pupils learning socially through discovery, exploration and making, as encapsulated in theoretical approaches to learning such as constructionism (Papert and Harel, 1991). In addition, learning theories like collaborativism (Harasim, 2017) apportion greater significance to socially constructed aspects of learning when using online platforms. Data can still be equally useful when promoting socially constructed learning. Pupils leave lots of qualitative data as they collaborate on digital activities; for example, this data could be posted on a discussion forum or shared blog. Data such as this offers the teacher insight that can be analysed to consider inclusive support strategies for individual learners.

Weegar and Pacis (2012) have offered useful discussion around the impact that online technologies have had on education at a theoretical level. They suggest that often a mixture of behaviourist and constructivist approaches are still deployed by teachers when using technology. Achieving a balance between the use of immediately impactful technologies and data that encourage positive pupil behaviours, alongside opportunities for more meaningful socially mediated discovery-based learning represents the most effective way to ensure an inclusive and well-balanced curriculum that engages all pupils.

Activity

Which technologies and apps could you use to promote pupil behaviours via "reward" and "punishment"? Which technologies and apps could you use to promote social and/or discovery-based learning? For both approaches, what data would you have access to for analysis?

Challenges to use of data in schools

Although we have discussed several specific tools that enable teachers to use data in productive ways that allow them to better know and support their learners, it is important to note that there are still practical, cultural and ethical challenges we must be mindful of when using data to support pupils in educational settings.

One technical issue relates to the siloed nature of the data that many of the applications we have discussed produce. For example, if a teacher requires a holistic overview of pupil progress via data, this would often involve the time intensive task of combining individual data sets from a range of individual applications, such as Teams, EdPuzzle and Kahoot. In addition to being time intensive, this requires teachers to have sufficient data literacy skills to combine different datasets practically. Many teachers often find themselves collating different data sets in their own spreadsheets; this can also create issues around data security and data loss with files stored locally on laptops, etc., rather than on secure platforms provided centrally by the school.

The Gates Foundations report (2015), "Making Data Work for Teachers and Pupils" surveyed 4600 and interviewed over 80 teachers in the United States who identified the below as the biggest challenges to data-driven practice. Much of this is still relevant within educational settings in the United Kingdom today.

The survey described the following issues that teachers had with data capture and utilisation in education:

- Overwhelming – large amounts of data from disparate sources make it challenging to separate the signal from the noise.

- Incompatible – require time-consuming manual data entry to put data to use.

- Inconsistent – in their ability to provide data reports in sufficient detail required.

- Too slow – to provide information in time to modify instruction in meaningful ways.

In the last ten years, the field of learning analytics has produced a swell of work in response to some of the barriers mentioned above.

Learning analytics is defined as "the measurement, collection, analysis and reporting of data about learners and their contexts, for purposes of understanding and optimising learning and the environments in which it occurs" (Siemens and Long, 2011).

Increasingly sophisticated learning analytics platforms are being utilised by schools, colleges and universities, which are capable of quickly collating diverse data sets into a single space to address many of the issues discussed in the Gates Foundation report. However, many teachers still find themselves at schools where there is not a central platform that neatly collates and displays varied datasets on their behalf.

Solutions like iDoceo on iOS offer a "one stop shop" where individual teachers can store varied data sets on their pupils in one location. For example, teachers can store Pupil grades, timetables, seating plans, resources and attendance data all in one location. This allows teachers to more easily manipulate and analyse data due to it being stored in a single location. Check out the iDoceo website at <u>idoceo.net</u> for further information and tutorials about the tool. You can also access dummy datasets to familiarise yourself with functionality before importing live pupil data as below.

Another issue we must acknowledge is less of a technical one, and more related to a cultural reluctance to embrace data amongst teachers. To overcome this issue, teachers and pupils should be involved in conversations around the data we need to capture and how these data should be used (Nesta, 2016). Increased transparency in this area will help teachers see data as something that can help them better support their pupils.

We must also ensure data-related success stories are celebrated with colleagues. For example, data analysis may have inspired a teacher to intervene successfully with a specific pupil, meaning they go on to achieve their potential in that subject.

Activity

What are your concerns about teachers using increasing amounts of data in their practice? What concerns might pupils also have about their learning being monitored via datasets? How can we overcome these challenges to ensure data can be harnessed to promote an inclusive curriculum?

Conclusion

This chapter should have enabled you to consider data from a more historically informed and broader perspective. Several technologies and educational scenarios have been explored to demonstrate the complex relationship that exists between the increasing use of digital data and the desire for an effective and inclusive education system. We have outlined some specific justifications and benefits that data provides teachers in supporting classes and planning for inclusion.

Think about the tools and applications you use in your teaching and the learning data that these records provide based on the examples provided in this chapter. Some of this data will no doubt be useful in improving your practice and better supporting your learners, via more nuanced formative assessment and informed differentiation. Equally, consider where excessive use of data can become unhelpful for teachers and/or pupils, hindering inclusion by labelling and categorising pupils inappropriately or unfairly.

This reading should have enabled you to consider some of the broader challenges and ethical issues that arise with increasing amounts of pupil behaviour being captured as data. Think about ways you can promote increased transparency with pupils, so they fully understand which of their activities are recorded as data, and for what purpose.

Pupils should have the opportunity to input into the decision-making process about how data are used to support them; additionally, they should have the space to enable them to contextualise and narrate their data with their teacher. Finally, this chapter should have allowed you to start considering the implications the acceleration of educational technologies and data-informed practice has for the educational sector at a practical, theoretical and ethical level.

By increasing teacher awareness in this space, this will help to ensure the potentials of data-informed practice can be realised, whilst at the same time minimising the extent of any practical and ethical challenges.

Annotated bibliography

IDE-Based Learning Analytics for Computing Education: A Process Model, Critical Review, and Research Agenda (Hundhausen et al., 2017) – this paper explores the potential benefits that data-informed insight can offer teachers in developing a picture of pupil learning during programming. A social media chat panel was implemented into the IDE for pupils to discuss their programs, generating further data for analysis. This enabled the teacher to develop a rounded picture of pupil's understanding and progress.

Learning Theory and Online Technologies (Harasim, 2017) – this book explores traditional and contemporary learning theories and the impact these approaches have had on digital and online education.

Making the most of data in schools (Nesta, 2016) – this report summarises the findings from a roundtable event involving leading actors from across the UK education sector. The opportunities that data offer teachers, as well as the barriers that still need to be overcome, are discussed.

References

Ahern, S.J. (2018) 'The potential and pitfalls of learning analytics as a tool for supporting pupil wellbeing.' *Journal of Learning and Teaching in Higher Education*, 1(2) pp. 165–172.
Bill & Melinda Gates Foundation. (2015) *Teachers know best: Making data work for teachers and pupils*. ERIC Clearinghouse. https://s3.amazonaws.com/edtech-production/reports/Gates-TeachersKnowBest-MakingDataWork.pdf
Brown, N.C. and Wilson, G. (2018) Ten quick tips for teaching programming. *PLoS Computational Biology*, 14(4), p. e1006023.
Coleman, V. (2021) Digital divide in UK education during COVID-19 pandemic: Literature review. In *Cambridge assessment research report*. Cambridge, UK: Cambridge Assessment.
Department For Education. [Online] [Accessed on 21st February 2021] https://www.gov.uk/data-protection
Harasim, L. (2017) *Learning theory and online technologies*. New York: Taylor & Francis.
Hundhausen, C.D. Olivares, D.M. and Carter, A.S. (2017) 'IDE-based learning analytics for computing education: A process model, critical review, and research agenda.' *ACM Transactions on Computing Education*, 17(3) pp. 1–26.
Nesta. (2016) *Making the most of data in schools*. [Online] [Accessed on 20th March 2021] https://media.nesta.org.uk/documents/Making_the_most_of_data_in_schools_-_FINAL.pdf

Ofsted. (2008) *Using data, improving schools.* [Online] [Accessed on 17th February 2021] http://www.educationengland.org.uk/documents/pdfs/2008-ofsted-using-data.pdf

Papert, S. and Harel, I. (1991) Situating constructionism. *Constructionism*, 36(2) pp. 1–11.

Price, T.W. Dong, Y. and Lipovac, D. (2017) iSnap: Towards intelligent tutoring in novice programming environments. In Proceedings of the 2017 ACM SIGCSE Technical Symposium on Computer Science Education, pp. 483–488. New York: Association for Computing Machinery.

Siemens, G. and Long, P. (2011) 'Penetrating the fog: Analytics in learning and education.' *EDUCAUSE Review*, 46(5), pp. 30.

Stephens, W.B. (1998) School attendance and literacy: 1750 to the later nineteenth century. In *Education in Britain, 1750–1914*. London: Palgrave, pp. 21–39.

Van Dijk, J. (2020) *The digital divide*. Cambridge: Polity Press.

Weegar, M.A. and Pacis, D. (2012) 'A comparison of two theories of learning-behaviorism and constructivism as applied to face-to-face and online learning.' In Proceedings of the 2012 E-Leader conference. Manila. https://d1wqtxts1xzle7.cloudfront.net/34484829/compare_of_two_teory-libre.PDF?1408476901=&response-content-disposition=inline%3B+filename%3DCompare_of_two_teory_PDF.pdf&Expires=1687177542&Signature=LN8BgeoJ1HD4AV0zpdEkxnIhZ2PP6swXPFdeBnPUrHq~H8gbDURgltGCX-i3AY3xUC9TP480rsEJTGR-tWSBUI2xn8c0XUfkzX0CgQrIxM07N1KWJwHBHNtfZ-BY7n-Mh43MrwAXp5HtBOQfyWIdFFQzZv3wzGHpLzA5WkHl2LeWGvGVQA~YfhpMB-1h9FmvBkoi0V9iRAEBnW07pZxyzmPTWSe6yshas3xxVXHYAOZjro0Gxq3tM-pNROzLp-WVUG6GVQDlNbbW3gP10KDQYvs72NGuULy9CwncxvUXWEULkI1x9zmC~5utdDQTsY-cYA56nyuuJgDUTJuMRHKsf-OZS4w__&Key-Pair-Id=APKAJLOHF5GGSLRBV4ZA

Yin, D. (2019) *Guide to actionable pupil data part 1 — How teachers should take actions with pupil data.* Alef Education. [Online] [Accessed on 24th February 2021] https://medium.com/alef-education/guide-to-actionable-pupil-data-part-1-how-teachers-should-take-actions-with-pupil-data-da20889dcbb5

Opting out

Why are pupils choosing not to study computing?

Cathy Lewin and Eleanor Overland

Introduction

The General Certificate of Education (GCSE) is the main type of qualification taken by pupils aged 16 in England. Pupils have compulsory core subjects, which include English, mathematics and science. In addition, pupils can select other subjects, often described as "option subjects". Typically, pupils select their options at the end of Year 9 (aged 13 or 14) and go on to study these subjects until the end of compulsory education at age 16. Annual GCSE data on qualification uptake provides a way to see the development of the subject over time and to make direct comparisons with other subjects. The more pupils who select computer science GCSE, the greater the number of pupils who are able to select it for further study such as A Level (Advanced Level), apprenticeships and degrees. Decisions made to select computer science as an option in Year 9 can therefore have huge implications for the size and nature of the future workforce in computing industries.

The "After the Reboot" report (The Royal Society, 2017) analysed the GCSE entries in computer science. The first concern was the size of the cohort; in 2017, only 11% of pupils were taking GCSE Computer Science. Of these, only 20% of the GCSE Computer Science cohort were female. Black minority pupils were underrepresented in the cohort. Chinese pupils and those from other Asian backgrounds were more likely to study GCSE Computer Science, although they are less likely to continue to study it beyond GCSE. Pupil premium pupils (a measure of disadvantage identifying pupils from least affluent backgrounds) were underrepresented in the subject, as were pupils with special educational needs or disabilities (SEND) (The Royal Society, 2017). GCSE data is published annually, and 2022 data from Ofqual Analytics (https://analytics.ofqual.gov.uk/) suggests that little has changed in the last five years.

The barriers to opting for GCSE Computer Science are numerous and complex. They include structural issues, such as the availability of specialist teachers. They also include identity issues, such as girls feeling that they don't belong in what

DOI: 10.4324/9781003193685-7

they see as a male dominated domain. In this chapter, we focus on the barriers that can arise in relation to the ways in which schools organise the Key Stage 3 curriculum and the GCSE options process. These affect both girls and marginalised groups, such as students from poorer backgrounds and some black and minority ethnic (BAME) groups, especially black students.

A major barrier is the availability of the subject at Key Stage 4. GCSE Computer Science is not offered at all in some schools. In 2018–2019, it was only offered in 61% of all schools (Kemp & Berry, 2019). Notably, the subject is offered at fewer girls' schools than boys' schools, although these schools are in the minority compared to mixed schools (Kemp & Berry, 2019). In addition, schools serving more disadvantaged communities are less likely to offer Computer Science GCSE (Kemp & Berry, 2019). The "After the reboot" report (The Royal Society, 2017) identified a shortage of specialised teachers as a barrier to the availability of GCSE Computer Science, with 44% of computing teachers feeling more confident in delivering the computing curriculum for younger age groups. Many schools, particularly in areas with wider teacher recruitment challenges, such as rural areas, find recruiting and retaining specialist computing staff difficult (The Royal Society, 2017; Worth & Van den Brande, 2019). A lack of awareness from parents and governors of the importance and relevance of this subject negatively impacts demand for the subject, resulting in GCSE Computer Science not becoming a recruitment priority for the schools. Pupils can therefore be stuck in a vicious circle involving a lack of demand and a lack of subject-specialist teachers.

Another structural issue can be the way the school timetables are structured. Schools predominantly present GCSE options in blocks (groups of subjects from which students must select one or two choices) to facilitate timetabling (Abrahams, 2018). This means that opportunities can vary from school to school, and young people's choices of GCSEs can be constrained (Abrahams, 2018; Barrance & Elwood, 2018). Often, it is schools serving the most disadvantaged that have the least flexibility and fewest resources to really offer students choice, offering a very limited range of subjects. This makes choice "classed", as argued by many (e.g. Abrahams, 2018; Henderson et al., 2018).

Another barrier can be one intentionally imposed by teachers themselves. Some schools also apply filtering in the options process, preventing students who are less academically able from taking up the subject (Barrance & Elwood, 2018; Cheryan et al., 2017; Mee, 2020). For example, some schools use mathematics attainment as a filter (Kemp and Berry, 2019). Other subjects perceived as hard are similarly restricted to high achievers (e.g. foreign languages). In addition, teacher guidance to students during the options process can actively reinforce stereotyping and gendering of subjects, even unconsciously through small asides (Gillborn, 1990), as can guidance from parents and siblings (Cheryan et al., 2017; Dos Santos et al., 2021; Wong & Kemp, 2018). Furthermore, disadvantaged students may have fewer resources and less support (less "cultural capital") for making decisions about their options (Barrance & Elwood, 2018; Henderson et al., 2018).

Finally, with some schools offering reduced curriculum time at Key Stage 3 and some offering computing as a carousel subject at Key Stage 3, with pupils only studying computer science for a few weeks every year, it can be very difficult for students to have the knowledge and experience to make informed choices about the subjects they would like to study at GCSE level. This, combined with the organisational factors listed above and the cultural factors discussed elsewhere in this book, results in GCSE Computer Science not being an inclusive subject.

The option process in schools is very difficult to plan. It provides school leaders with complex issues around staffing and curriculum time. These organisational aspects of the options process can significantly impact how young people experience the process and their decisions. This chapter explores research on how young people experience the GCSE options process (completed pre-pandemic) and explores potential actions schools could take in order to increase future numbers of underrepresented groups. We also consider how schools can use online and face-to-face activities to provide creative and collaborative approaches to supporting the options process in the future.

What the research found

In order to explore some of the stories behind the statistics, we undertook two case studies, gathering data from staff and students. We spent time in two contrasting schools in North West England. We surveyed all pupils after they had taken their options and interviewed a selection (both those who had and those who had not opted for GCSE Computer Science). We also interviewed the heads of department and teachers and explored the options and information materials. In addition to surveys, the pupils also completed mind maps and drawing exercises to explore their wider perceptions of computing.

The data were analysed through a sociocultural view of identity and agency, drawing on Holland and colleagues' (2001) figured worlds. Figured worlds views identity as fluid rather than fixed, with beliefs and behaviour strongly influenced by cultural models and narratives. Here, we draw on two main concepts: positionality and authoritative voices. People in figured worlds position themselves and are positioned by others. For example, a student may position themselves as good at computer science. Students are not passive recipients of positioning but may accommodate, deny or resist invitations by others to assume particular positions. Authoritative voices (or discourses) are external and persuasive. Using this approach allowed us to have a more nuanced understanding of how young people position themselves and are positioned by others, such as teachers, their families and peers.

Whilst both schools had secured healthy numbers for their GCSE Computer Science classes, they would have liked to recruit more. Both schools had far more boys than girls opting for Computer Science GCSEs (in line with the national trend).

At the time of this study, school A operated a two-year Key Stage 3, with students making choices about which GCSEs to study in the spring term of Year 8, taking nine subjects. GCSE Computer Science was offered as one of five English Baccalaureate (EBacc) subjects but could also be chosen as one of three other options (effectively four opportunities to select computer science). In Year 8, prior to the options process, units relating to both computer science (e.g. programming) and ICT (e.g. website design) were taught, with the subject area being highlighted to students so that it could inform their decision-making. Take-up of the subject had expanded, with two groups starting GCSE in 2017–2018 as compared to one group previously. Of the 84 students who responded to the survey, 18.3% opted for GCSE Computer Science (including one girl), 13.4% opted for both the GCSE and a vocational ICT course, 9.8% opted for a vocational ICT course only (including three girls), whilst 58.5% did not choose computer science or vocational ICT.

At school B, students were expected to take at least eight subjects. Students were expected to choose between computing (a vocational course) or separate science as part of the core curriculum. GCSE Computer Science could be selected as "one free choice of any subject" (effectively one opportunity to select computer science), and so was in competition with many other courses. In the lead up to submitting their options choices, the computer science teachers talked about the subject, showed the students exam papers, gave a presentation in assembly, and welcomed current students (a boy and a girl) into the classroom to talk to students going through the options process. Once students had submitted their option choices, they had an interview with a member of the senior leadership team. ICT and computing were taught for one hour a week in Years 7, 8 and 9, prior to students choosing their options. Of the 158 students who responded to the survey, 15.2% chose GCSE Computer Science (including five girls). Data were not collected on the students taking the vocational information technology-based course.

The positioning of computer science through options

In line with the literature summarised above (e.g. Abrahams, 2018), some pupils found the "option blocks" prevented them from being able to select computer science as it was set against some of their other favourite subjects. One girl, at School B, faced with one option box from which she could choose computer science said:

> I really want to do art but really want to do computer science, so I chose art. [...] It was hard to choose between both of them. [...] It was really difficult and stressful as well [...].

Others talked about the EBacc requirement, university applications and how GCSE Computer Science was not perceived to contribute towards meeting these. One boy, at School B, said:

> I really thought I'd have a better opportunity at getting somewhere with history, PE and French. With the English Baccalaureate you needed a language to actually do it.

GCSE Computer Science is part of the EBacc under "sciences". Given that schools usually offer combined science (a double award at GCSE) or separate sciences (biology, chemistry and physics) there is no reason for students to opt for computer science to meet the EBacc requirements. Students also commented on the need to balance workload, noting that choosing triple science alongside computer science would not be manageable.

Filtering was an issue in School B, which focussed especially on pupils in higher-ability mathematics sets, although it did consider less able students. The survey in this school identified a high number of pupils who did not take computer science as they felt it would be too difficult for them. One teacher explained:

> So sometimes we'll look at their maths abilities. So if they're quite good at maths, this would be a good option for them. We don't restrict the others, saying, "You're not capable" or "You can't do it." But that does kind of influence us to push the smarter ones, the brighter ones, but we don't say to the weaker ones, "You can't do it." But we probably promote it more to the smarter kids.

The students were told in assembly that being good at maths would be helpful. Computer science was also positioned as a difficult subject by teachers, who expressed this opinion to both students and their parents. Similarly, one member of staff noted that:

> Nobody knows what happens in [the interview with a member of the senior leadership team], how much of it is student choice and how much of it is guidance. […] even though you want the child to pick something they enjoy the results will always come first and that will always be the influencing factor.

That is, from a school perspective, lower ability students were positioned as not being able to achieve a "good" grade in computer science and steered towards other subjects.

The influence of authoritative voices

As part of the options process in both schools, students were encouraged to listen to what teachers had to say and to discuss their option choices with family and friends. This affects how students position themselves in relation to the subject of computer science, that is, whether or not they see themselves as someone who can take Computer Science GCSE.

The students in the research were highly influenced by the advice and experience of those around them, particularly older siblings or other family members. One female student (from School A) who had opted to take GCSE Computer Science explained how her brother had strongly influenced how she positioned herself in relation to computer science. He had studied computer science at college and had been offered an apprenticeship:

> and when I heard he got that and stuff, it just wowed me [...] I've liked computer science for a long time because I love working with computers, and since he did that, I was like, "This is what I want to do, I really want to get into this." So, from there, I started choosing it.

Through her interview, she positioned herself as good at computer science, and comments from her teacher and brother contributed to this perception. Similarly, another girl from School B said that her brothers and sisters had also chosen the subject and that her father was a computer science professional.

As well as siblings, teachers also play an important role. For example, a girl (from School B) cited a "persuasive" assembly by the Head of Department alongside guidance from her older sisters, who did some research on the subject on her behalf. A boy at school A explained that he was going to choose the vocational ICT course, but his teacher told him that he was good at computer science and could do the GCSE, so he changed his mind. He was able to re-position himself as someone good at computer science. A teacher from school A also said: "I have even managed to take two boys out of triple science, to convert to me". This suggests that the teacher had worked hard to persuade the boys to drop triple science in favour of computer science.

Similarly, teachers' opinions and their assessment data could position students as "not good enough" at computer science to take it at GCSE (particularly in school B, where filtering was in place). Some students found it hard to resist this authoritative discourse, rapidly ruling computer science out. The survey data suggest that nearly half of students from both schools were influenced by their teacher either a lot (16%) or a little (33%) when deciding whether or not to choose GCSE Computer Science.

Other aspects affecting how students position themselves in relation to computer science

Career plans, how students positioned themselves in the future workplace, played an important part in the decision-making process. Many students had already identified their preferred career (commonly teacher, lawyer and doctor) and did not see computing as relevant to their career choice. One girl explained how she planned to set up her own fashion company but would employ others to do the technology side of things. Later in the same interview, when asked about her option choices, she said:

> It wasn't shouting my name, I guess. I'm into fashion so I didn't think that IT or computer science would be related to anything I want to do in the future.

Gender was a particularly complex aspect to explore. Many pupils seemed conscious of describing computer science as a subject open to both boys and girls, but they also acknowledged that a lot more boys than girls chose it as a subject. The majority explained this with the factors already discussed. Some did specify gender as a feature:

> Most people are known for girls to do either PE or some dance or drama and stuff like that; and boys are mainly into the mathematical, science side of it, and computing, than girls.

Both schools in the project were highly multicultural, and some pupils explained how gender expectations may be more of a feature in some cultures and families.

Some pupils linked gender discussions to ability, with one female explaining, "Computer scientists are very smart, or they could be an average person who knows their stuff and is into it". She goes on to explain that boys are more likely to play computer games and be "in to computers", so they could therefore take it as an option even if they were "average", whereas the girls needed to be "very smart".

What can teachers do?

This research only focussed on two schools. However, it is interesting to see the similarities and differences between them. Following on from our discussions with the pupils and teachers, we have put together some recommendations you may wish to take forward in your own schools. Some perceptions are that the lack of diversity in computing is a societal issue rather than something that teachers and leaders can change within their own setting. There are many wider issues that do need to be addressed; however, opportunities to adjust structures, provide pupils with suitable authoritative voices and raise self-belief should not be dismissed.

Explore your own data: Does your school mirror some of the issues outlined in the national data? Is there a gender gap? Are disadvantaged pupils or those with special educational needs represented proportionately? Are your own option groups diverse? Scrutinising your own data for your computing classes will let you see exactly how inclusive the subject is in your school. It may be that you identify specific underrepresented groups that you wish to target specifically.

Carry out your own pupil survey: Use a quick online survey to find out why current KS4 pupils have and haven't opted for GCSE. It may reveal reasons specific to your setting and ones that may not have been identified in this research. You can then plan specific actions to support your own pupils.

Ensure Key Stage 3 creates readiness for GCSE Computer Science. This is particularly important where schools have computing as a carousel subject or are reliant on non-specialist staff to deliver at KS3. They may not be able to remember key knowledge from computer science and may feel unprepared to continue studying the subject further. It may be useful to review the curriculum, particularly in the run-up to options time, and ensure these lessons are engaging and prepare pupils for GCSE. The use of online materials may be an opportunity to revisit key knowledge. Where specialist staffing is in short supply, it may be appropriate to move specialised teachers to teach targeted groups in the run-up to options.

Give pupils self-belief: Have you decided within your school if GCSE Computer Science is open to all abilities of pupils? If you are selective and emphasise how difficult computer science is as a subject, you may be putting off other pupils who are more than capable of success. There are lots of ideas and resources available to encourage resilience and self-belief in computing, which it may be useful to explore, especially to promote more girls into STEM-related subjects. The WISE Campaign (2022) "My Skills, My Life" is a useful starting point.

Review option structures: Are pupils being forced to choose between computer science and another of their favourite subjects? Are pupils following the EBacc requirements unable to opt for computer science? Perhaps just moving subjects into different option blocks could increase pupil numbers. This is something you may want to look at and discuss with the leadership team, governors and parents.

Make career education a regular feature of Key Stage 3 computing: Pupils seeing the relevance of the subject for their future career is essential in encouraging them to continue to study it. Regular signposting to computing jobs and highlighting the contribution of computing to other professions can support pupils' understanding. Display materials, and guest speakers from industry, alumni and parents working in computing can really help with this. Chapter 10 of this book explores in more detail how you can collaborate with employers in the sector.

Use online materials to supplement options information: Curriculum time for Key Stage 3 computing classes can be limited, and it can be a challenge to fit in all the relevant activities and resources you may want pupils to access before they select their options. There are lots of ways you can engage pupils online or via extracurricular sessions to encourage them to consider GCSE Computer Science as an option. Host exciting computing competitions or projects for pupils to get them involved prior to taking their options. Ask current GCSE students and alumni to make videos about what they enjoyed in their computer science course. Create an online gallery (slideshare or similar) of GCSE class work and projects. Create a presentation targeted at parents/carers to provide information about computer science as a subject, opportunities for further study and careers. Provide links to computing careers websites, especially those that promote a broader representation of people. Ask local employers and higher education providers to tell pupils why computer science is important and the opportunities available (perhaps you could put together a panel and host an online computing careers event).

Share your success: Increasing numbers and increasing diversity are challenges for many computing teachers. A range of strategies and events are being tried in many schools. Some schools have had more success than others. Sharing any successes, even small ones, can support developing collective intelligence. It is also great to share resources and industry contacts. Local networks can help you collaborate. "How to use this book" at the beginning of this book helps you build your local network. Working with local teachers can help you build a more detailed picture of some, potentially localised, barriers to inclusion in computing. Most importantly, it can be a way to discuss ideas, develop collaborative opportunities and make a difference.

References

Issues identified in this report have recently been discussed in an excellent Raspberry Pi research seminar presented by Dr Peter Kemp and Dr Billy Wong (Wong B & Kemp P, 2018), which is available to watch online (https://www.raspberrypi.org/computing-education-research-online-seminars/previous-seminars/#underrepresented-groups).

Abrahams, J. (2018). 'Option Blocks that Block Options: Exploring Inequalities in GCSE and A Level Options in England'. *British Journal of Sociology of Education*, 39(8), pp. 1143–1159. DOI: 10.1080/01425692.2018.1483821

Barrance, R., and Elwood, J., (2018). 'Inequalities and the Curriculum: Young People's Views on Choice and Fairness Through Their Experiences of Curriculum as Examination Specifications at GCSE'. *Oxford Review of Education*, 44(1), pp. 19–36. DOI: 10.1080/03054985.2018.1409964

Cheryan, S., Ziegler, S.A., Montoya, A.K. and Jiang L. (2017). 'Why are Some STEM Fields More Gender Balanced than Others?' *Psychological Bulletin*, 143(1), pp. 1–35. DOI: 10.1037/bul0000052

Dos Santos, E. D., Albahari, A., Díaz. S. and De Freitas, E.C. (2021). 'Science and Technology as Feminine': Raising Awareness About and Reducing the Gender Gap in STEM Careers'. *Journal of Gender Studies*, 31(4), pp. 505–518. DOI: 10.1080/09589236.2021.1922272

Gillborn, D. (1990). 'Sexism and Curricular 'Choice'. *Cambridge Journal of Education*, 20(2), pp. 161–174. DOI: 10.1080/0305764900200207

Henderson, M., Sullivan, A., Anders, J. and Moulton, V. (2018). 'Social Class, Gender and Ethnic Differences in Subjects Taken at Age 14'. *The Curriculum Journal*, 29(3), pp. 298–318. DOI: 10.1080/09585176.2017.1406810

Holland, D., Lachicotte, W., Skinner, D. and Cain, C. (2001). *Identity and agency in cultural worlds*. Cambridge, MA: Harvard University Press.

Kemp, P.E.J. and Berry, M.G. (2019). *The Roehampton annual computing education report: pre-release snapshot from 2018*. London: University of Roehampton. https://www.bcs.org/media/2520/tracer-2018.pdf

Kemp P and Wong B (2021) Computing education for underrepresented groups. Raspberry Pi Foundation. Online [https://www.raspberrypi.org/computing-education-research-online-seminars/previous-seminars/#underrepresented-groups]

Mee, A. (2020). *Computing in the school curriculum: a survey of 100 teachers*. DOI: 10.13140/RG.2.2.19883.59689.

The Royal Society (2017). *After the reboot: Computing education in schools*. London, UK: The Royal Society. https://royalsociety.org/-/media/policy/projects/computing-education/computing-education-report.pdf

WISE Campaign (2022). My Skills My Life. https://www.wisecampaign.org.uk/my-skills-my-life/

Wong B & Kemp P (2018) Technical boys and creative girls: the career aspirations of digitally skilled youths, Cambridge Journal of Education, 48:3, 301-316, DOI: 10.1080/0305764X.2017.1325443

Worth, J. & Van den Brande, J. (2019). *Retaining science, mathematics and computing teachers*. Slough: NFER. https://www.nfer.ac.uk/media/3784/retaining_science_mathematics_and_computing_teachers.pdf

Gender differences in computing classrooms

Practices to develop inclusive learning spaces

Louise Hayes

Introduction

A gender imbalance exists at school, university and in the workplace in the field of computer science (CS). Whilst much research has been carried out to find out the reasons for the imbalance, the impact of classroom environments on the uptake of girls in the subject remains an under-explored area of enquiry. This chapter begins by looking at approaches to supporting girls in the subject, particularly focussing on pedagogy to build confidence and encourage interest in the subject. Then it considers how wall displays and classroom layouts can be better utilised to inspire girls in the subject. It draws on research and strategies that I used as a secondary school computing teacher and now as one of a few female computing teacher educators in England. This chapter is a result of these reflections and observations in response to my own experiences in school and, more recently, through training computing student teachers. As a new teacher in the subject of computing, you will find that you face specific challenges that are pertinent to the subject. Here, it is important that you reflect on your own learning preferences and your experiences from school, university and/or the workplace. You might want to think about who inspired you throughout your education, or what the gender balance was like in your experience of computing education.

Pedagogy impact on gender

In earlier chapters, we looked in more detail at the different approaches to teaching programming. This is particularly important when teaching girls, as research suggests this can have a detrimental effect on their learning. Different methods of programming pedagogy align with a number of perspectives, such as Piagetian-based constructivism, based on the work of Papert, where pupils gain computing skills and knowledge through leaders facilitating experimentation. Alternatively, a more scaffolded approach to learning programming aligns with Vygotsky's.

Here we consider the aspects of pedagogy that impact girls, and their interest and confidence in computing.

> Although pairing helps all students, we believe that it is particularly beneficial for women because it addresses several significant factors that limit women's participation in computer science. We provide reasons for our belief that pair-programming helps women persist in these majors.
>
> (Werner et al., 2004)

Studies have attempted to determine whether "creating a culture of collaboration would increase retention of women". Werner and her colleagues studied pair programming among university students; paired programming is where one is the driver who works the keyboard and mouse while the other navigates. Werner found that "successful use of pair programming requires collaboration rather than competitiveness and dominance" Werner et al. (2004). Raspberry Pi Foundation Pedagogy Quick Read supports the evidence that shows pair programming can benefit girls in terms of results and their perception of the subject, there is no evidence to suggest that it has a negative impact on boys (Childs K, 2022).

Having an understanding of the impact of pedagogical methods on gender is important for computer teacher educators. These methods are one of the factors that have an impact on subject uptake and continued study by girls. You need to ensure that you are adapting your lessons for all learners in your class, and as computing educators, ensure that you plan for gender and not just pupil ability. What you are aiming for is a lesson that meets the needs of all learners by adapting your planning. The impact of different pedagogies on girls, in particular, needs to be addressed in both the theoretical and programming aspects of your lessons to change classroom cultures that are not currently working. It is important that you know about and use the pedagogical methods that suit girls as well as boys in your lessons, even if you are teaching in a single-sex school.

Recognised as a foundational competency for being an informed citizen, computational thinking encompasses artificial intelligence, big data, speech and facial recognition. These are a few of the many aspects that are changing how we work, collaborate, and communicate (Grover and Pea, 2017). The English computing curriculum states all pupils must "have the opportunity to study aspects of information technology and CS at sufficient depth to allow them to progress to higher levels of study or to a professional career" (Department for Education, 2014). In 2018, a report by Roehampton University showed that the percentage of all students who opt to take General Certificate of Education (GCSE) CS increased marginally from 12.1% in 2017 to 12.4% of all GCSE students in 2018. Of these, only one in five (19%) girls in English comprehensive schools took the subject (Kemp and Berry, 2019). In a more recent report by the British Computer Society, we see that this figure has worsened: "girls are now outnumbered six to one by boys across computer science"

(The British Computer Society, 2022). This imbalance clearly continues to be a concern if this is seen as a skill for everyone, not just computer scientists, to learn.

Acknowledging that not everyone will be, or wants to be, a computer scientist, it is vitally important that this situation be changed. A reflection action here is for you to consider how innovative practises might be embedded both inside and outside of your classroom to encourage the uptake of girls. A school timetable and option blocks. If girls are not able to access the subject due to this reason, as seen in Chapter 6, how might you work around this? You may need to seek support from your senior management or leadership team within the school to do this. Instigating change at the classroom practitioner level comes with its own set of challenges, but as computing educators, it is vitally important that we have the tools to hand to change existing practises and support girls in the subject.

Relevant role models

In a workshop I run at the university, the PGCE computing student teachers are asked to create a gender-relevant wall display. This often opens up a strongly opinionated debate, and students are often perplexed as to why they are asked to do this. The students are asked to look at the walls in their training classrooms. It is certainly not an activity designed to criticise classroom practises; rather, it reflects on how they might positively impact gender and inclusion. They are asked to take pictures of their classroom and plan a wall display. The image of a computing classroom wall display (Figure 7.1) shows a number of CS-related careers. The list is a good starting place to think about careers, and you can think of what you might add to it. The labelling shows names of job titles, such as "forensic

Figure 7.1 A secondary school careers wall display. Photograph by the author.

computer analyst", "data analyst" and "web designer". Here, we might want to consider what they actually do and include relevant images to support the titles.

Underpinned by theories of motivation, expectancy-value theory identifies parents' and society's attitudes and perceptions towards the subject, where value is placed on the subject or career. Eccles highlights that "students are more likely to choose computing if they believe they will succeed and if they have a sense of support from those around them" (Eccles et al., cited in Dee & Gershenson, 2017). Addressing computing classroom environments and wall displays will support girls in your classes. Role model research examining the masculine stereotype of peers in CS has found that women prefer female role models, and less stereotypical role models (Master et al., 2021). The aim is for you to reflect on how you support girl's succeed in the subject and build confidence in your classes. How are they able to see and access role models that will inspire them? As a secondary school computing teacher, I planned a number of enrichment visits, some that were one-day and others that were longer visits abroad. The purpose of this was to understand industry practices linked to the curriculum. One visit was to a local BT Telephone exchange, and another a trip to Silicon Valley in the USA, which I used to build up contacts with local industry and universities. If this is not possible for you to do, another area for you to explore is your local CoderDojos and STEM Ambassadors, who will give you access to these contacts as they are often industry volunteers themselves (STEM Ambassadors, n.d.). Where schools work in partnership with local industry and build contacts, there will be the opportunity to connect or help develop connections outside of the classroom.

Planning a lesson

- Lesson planning that includes a collaborative group activity to produce a wall display that shows the stories behind the job titles. The objective is to show the involvement of women in industry and provide a detailed description of what they do. For example, plan a lesson that has the objective of producing a historical timeline for use as a wall display. An outcome of this is to create a sense of belonging by seeing some of the roles and key figures and raise aspiration of pupils. This could include job titles and descriptions using examples taken from the industry and images that are diverse in roles and people. You may want to include relevant films that can be included as part of lessons or as a homework task – there are some great inspirational stories to tell in the subject!

- Raise the profile of the department and work with other subjects on a cross-curricular activity that uses technology in the project work. The use of technology to support collaboration has increased and schools are using platforms to continue to support teaching. Here you can ask the pupils to use the share option to create a joint piece of work and share it across the platform for the school (a virtual wall display).

- Pupils can create a timeline of competitions and events in the annual calendar that is displayed on the classroom walls and around the school in digital displays. Assign digital leaders to lead the project – reflect on the data to see who might lead a digital group online for the task. Reflect on the activities that the pupils might support that are gender-relevant to them as part of this activity. You could have a regular calendar of events that are advertised on the department calendar, the learning platform and/or on your classroom walls.

- Wall displays that include free, downloadable, set of posters; one such excellent resource has been developed by CC4FN with a specific focus on "to normalise the fact that computer scientists come in all varieties. There are people of different gender, ages and ethnicities in the poster set" (Queen Mary University). This set of free resources can be found at the Queen Mary University (CCF4N) (Queen Mary University, n.d.).

Computing classroom designs

Supporting pupils to achieve and succeed is clearly the aim of any teacher. Research into all aspects of classroom design and even the "optimal amount of wall adornments" (Cheryan et al., 2014) that support maximum pupil achievement can be further explored. For the purpose of this chapter, we will consider computing classrooms and how they might be better arranged. Cheryan et al. (2014) found that stereotypical objects steer women away from CS and signalled who "belonged" in the space. If the layout and design of the computing classroom are considered off-putting to girls and therefore one of the reasons for them not taking the subject, how might you make some simple changes in the space? These could include "plants and inspirational posters, and reorganized materials to make them more easily accessible" (Cheryan et al., 2014). Computing classrooms come in many shapes and sizes, and you may find that they are often rooms that have been adapted to accommodate the equipment, where computers are sometimes an afterthought in the design.

Figures 7.2 and 7.3 show two examples drawn by two PGCE computing students in an activity where they were asked to reflect on their favourite classroom

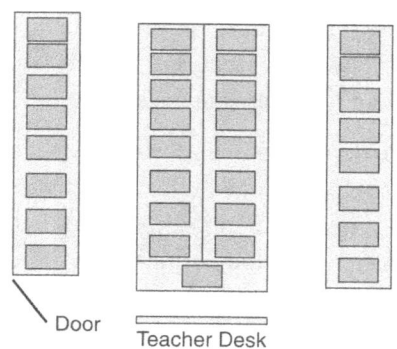

Figure 7.2 A secondary school computer classroom 1. Image by the author.

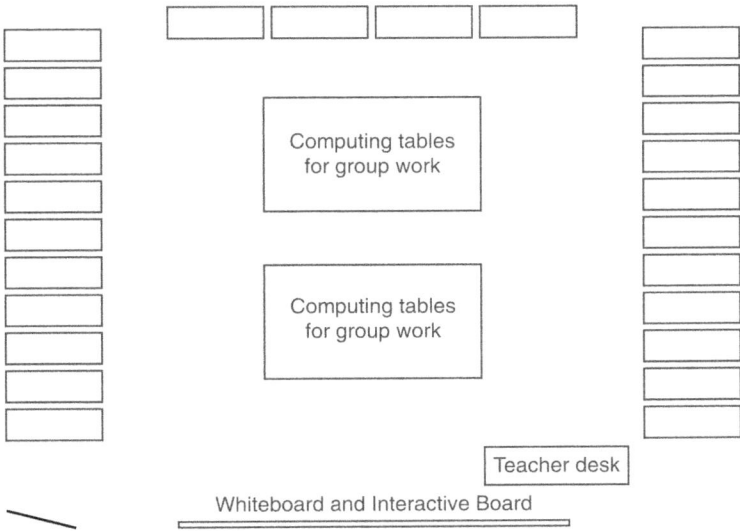

Figure 7.3 A secondary school computer classroom 2. Image by the author.

Activity

A useful exercise for you to do is to consider the layout of your classroom and examine how you will use the environment to encourage and support girls in the lesson. Do you have an area in your classroom with a selection of books and magazines (e.g. Hello World magazine: https://helloworld.raspberrypi.org/) that encourage wider reading and interest in the subject? Can you move the desks to enable groups or pairs to collaborate? Using the layout as a guide, where will you sit the girls/boys in the room? Have you asked the girls/boys what they like or don't like about the space?

In applying collaborative practises in your teaching, you might want to think about how you will plan theory and programming for group and paired activities into your lessons. The aim of this is to build confidence at an early stage so that girls continue with the subject. Here, you might want to think about the age at which you start doing this. Do you put your Year 7 pupils into pairs to support and build their confidence? If you are teaching in a classroom that is difficult for group work, you may even want to consider moving to a classroom that has no computers in it. Whilst not directly linked to classroom layouts, this is interrelated, as the layout needs to enable collaborative practise. There are quite a number of questions for you to reflect upon here.

Anecdotally, I have often noticed that girls will sit at the computers on the wall nearest to the door if left to their own devices. This was the case when I was a schoolteacher, and it is still the case in the sessions that I observe with my PGCE students. Take a look and see if this is the case in your classroom if pupils are left to sit where they choose!

design for teaching both the theoretical and practical aspects of the subject. As part of the session at the university, the computing education students reflect upon the differences in planning for teaching the differing aspects of the subject as specified in the GCSE CS specification, which include computer systems, computational thinking, algorithms and programming. They are asked to use the Raspberry Pi Foundation Pedagogy Quick Read "Pair programming supports learners to produce better solutions to complex programming problems", which also supports the planning of a subject pedagogy assignment as part of the course. The student teachers compare their own classrooms and contrast the designs. A further aim of this is to reflect upon the planning and adaption of teaching for gender pairs in their lessons.

Using data to further adapt planning

The rationale is for you to reflect on your own classes and consider how the subject-specific element of teaching computing impacts on your classroom. There is no one-size-fits-all solution, and each school and class will have a different set of needs. This is where you can more effectively use assessment data to help with your planning. Collaboration and peer learning can be supported by carefully constructing assessments, such as using Kahoot, Eedi or Teams Forms (Chapter 5 gives more in-depth detail on this). The aim is for you to consider how, through the better use of assessment data, you are able to further support girls in your lesson planning. This is an activity that you might want to reflect upon that will support building confidence in girls. In your planning, have you included the dates of birth of the pupils in the data set? Reflect on how often you hand out positive rewards to the girls in your class? You are able to use an app, such as https://www.idoceo.net/index.php/en/ technology to help you record this (which is also helpful for parents' evenings).

A further point to note here, which is a specific issue for computing teachers, is that you are very likely to have a lot of classes to plan for and assess on a weekly basis if you teach Key Stage 3 and Key Stage 4. At Key Stage 3, whilst you may only see the pupils once a week, you will need to get to know your classes well to enable accurate assessments of their progress.

Whilst the focus of this chapter very much sits within the lens of gender, certainly as a computing educator, I have always found that a great benefit of having digital skills is having the ability to use the technology to my own benefit as a practitioner. Because of this, you will find that you are able to easily use technology, devise assessments that inform your planning and manipulate the outputs to provide specific data sets. There are many apps and websites for baseline testing and assessment (Chapter 5 gives more information here), but do remember to check that they comply with any school policy, or if there is not a policy for this, discuss it with the relevant member of staff the need to adapt or create one.

This will not only help you to accurately plan but also ensure you capture all of the correct data in the first place to avoid GIGO ("garbage in, garbage out" as mentioned in Chapter 4).

Scenarios: In this final section, we will consider two examples taken from PGCE Secondary computing student lesson plans. Reflect on the questions in the section above to consider how you might use data to adapt your planning to the given lesson scenarios. In the activity below, the student teachers are asked to consider the example scenarios. They take part in a collaborative planning activity where they discuss how they might adapt the lesson's planning and compare the scenarios to the pupils in their classes (Table 7.1).

Table 7.1 Student–teacher lesson scenarios

Zara's classroom
Lesson Topic: 3.8 Create, Reuse, Revise and Repurpose Digital Artefacts for a Given Audience, With Attention to Trustworthiness, Design and Usability.

Pupils: Year 7 – The pupils are of mixed ability and balanced in gender; 17 are boys and 15 are girls.

Demographics: date of birth, gender, personal and subject interests.

Classroom design: Figure 7.2

Consider how you might group pupils using the demographic data for Zara's Year 7 class. How might you seat pupils if you decide to put them in pairs or groups? Will literacy, numeracy and age inform your seating plan? Or might you consider data on ability, including the digital literacy skills of the pupils?

Lesson activity: The pupils can be asked to create a blog or an poster where they collaborate to produce a themed activity that is gender and culturally relevant to them using technology. This supports your achievement of Teacher Standard 3, 4 and 5 (DfE, 2011).

Discussion:
- Reflect on the management of this activity and how you will arrange the pupils in the space.
- Reflect on how you will include in your lesson planning the details of appropriate use of language and questioning to ensure that responses are appropriate and considerate of others.
- How will you ensure that they are collecting relevant information?
- How might you support the literacy levels of your pupils in their web searches? For example, the website Choosito https://www.choosito.com/ is a smart search engine that safely searches the web for resources based on your students' reading level (Choosito, n.d.).

(Continued)

Table 7.1 (*Continued*)

Aaron's classroom

Lesson Topic: GCSE AI The Ethics of Artificial Intelligence.

Pupils: Year 10 computer science examination students in a high ability setting; there are 28 pupils, 4 of whom are girls.

Classroom Design: Figure 7.3

Question to consider: Take a moment to consider the following questions from the GCSE examination.

BBC Bitesize Example essay-style question about artificial intelligence

A company is considering buying a program that uses artificial intelligence to answer technical support calls. If the machine cannot answer the question, the customer will be passed on to the technical support workers. Discuss the implications of this technology.

The example has been taken from an AQA past paper. It reflects the type of question that may appear in an exam paper. BBC Bitesize

Here are some ideas for you to reflect upon on how to reduce gender bias in our teaching practice and curriculum using the examination example above:

- How might the question be adapted to balance gender examples? Use areas of interest to adapt the questions; for example, ask the pupils to research specific examples of its use in the real world.
- Watch the YouTube clip to support a class discussion to support the understanding of the perspectives of others from both a boy's and girl's perspective – LikeAGirl https://www.youtube.com/watch?v=XjJQBjWYDTs
- Compare past papers and examination boards to compare how this question has been set. How might questions be adapted to be critical of the limitations or mistakes in the development of technology? What examples are used in the question, and how might they be adapted to be relevant to the pupils?
- How will you create a set of ground rules for discussions about topics? Are you ensuring that language and behaviour are appropriate? How will it be moderated for "fake news" (the website Choosito might support you here)?
- Are you helping all of the pupils in your care? Have you collected and used assessment data to pair the pupils based on personality and interests rather than ability? Have you ensured that the assessments are for paired, group and individual work?
- How are you encouraging the pupils to work collaboratively on group or paired tasks? Do you encourage all of your pupils equally in your lessons? How are you supporting "self-efficacy" and encouraging confidence in the quieter girls in your care?

Concluding thoughts

In summary of this chapter, I hope that it has given a guide of how computing classrooms can be more engaging for lessons. A future more in depth piece of work or research might include a discussion that includes how digital skills might be developed for industry, the impact of artificial intelligence in computing education or a more research driven perspective on pedagogical methods to support inclusive classroom practices. But for now, I return to the main aims of this chapter, where I have asked you to consider what you might do to challenge and change practices in your classroom to support girls. In acknowledging that the issue is more complex than simply changing classroom environments and wall displays, the aim of the chapter was to ask you to reflect on adapting teaching and making classrooms engaging to consider gender. Using collaboration, wall displays that are aspirational and to motivate girls, support their confidence, as well as developing interests in the subject, will hopefully all have an impact.

To finish this chapter, I recommend that further reading on inclusive teaching practices, highlighted earlier in this work, will help and support your achievement of Teacher Standard 5 (DFE, 2011). I wish you good luck in helping to raise the aspiration and interest of all pupils to understand the benefits that computing and digital skills can bring to future careers.

References

BBC Bitesize. https://www.bbc.co.uk/bitesize/guides/zmxxh39/revision/3 [Online, Accessed 25 January 2022]

Cheryan S., Ziegler S.A, Plaut V.C, & Meltzoff A.N. (2014). 'Designing Classrooms to Maximize Student Achievement'. *Policy Insights from the Behavioral and Brain Sciences*, 1(1), pp. 4–12. doi:10.1177/2372732214548677

Childs K (2022) A pair programming approach for engaging girls in the computing classroom: study results. Raspberry Pi Foundation. Online [https://www.raspberrypi.org/blog/gender-balance-in-computing-pair-programming-approach-engaging-girls/]

Choosito. (n.d.) Search and Learn. https://www.choosito.com/ [Online, Accessed 25 February 2022]

Dee, T., & Gershenson, S. (2017). Unconscious Bias in the Classroom: Evidence and Opportunities. Google's Computer Science Education Research, Department for Education. https://assets.publishing.service.gov.uk/government/uploads/system/uploads/attachment_data/file/1040274/Teachers__Standards_Dec_2021.pdf

DfE. (2011). Teachers Standards. https://www.gov.uk/government/publications/teachers-standards https://assets.publishing.service.gov.uk/government/uploads/system/uploads/attachment_data/file/239067/SECONDARYnationalcurriculum-Computing.pdf.

DfE. (2014). National Curriculum – Computing Programmes of Study: Key Stages 3 and 4. https://assets.publishing.service.gov.uk/government/uploads/system/uploads/attachment_data/file/239067/SECONDARY_national_curriculum_-_Computing.pdf.

Coderdojo (N.D.). The Community of 2337 Free, Open and Local Programming Clubs for Young People. https://coderdojo.com/

Grover, S. & Pea, R. (2017). *Computational Thinking: A Competency Whose Time Has Come.* doi:10.5040/9781350057142.ch-003

iDoceo. Teacher Gradebook App. https://www.idoceo.net/index.php/en/

Kemp, P.E.J. & Berry, M.G. (2019). *The Roehampton Annual Computing Education Report from 2018.* London: University of Roehampton.

Master, A., Meltzoff, A. N. & Cheryan S. (2021). *Gender Stereotypes About Interests Start Early and Cause Gender Disparities in Computer Science And Engineering.* Published by PNAS. https://doi.org/10.1073/pnas.2100030118

Ofsted. (2022). Research review series. https://www.gov.uk/government/publications/research-review-series-computing/research-review-series-computing

Queen Mary University of London. *Computer Science for Fun* (CS4N). http://www.cs4fn.org/art/hideandsee.php [Online, Accessed 21 January 2022]

STEM Ambassadors. STEM Ambassador Programme. https://www.stem.org.uk/stem-ambassadors

Werner, L. L. Hanks, B., & McDowell C. (2004). Pair-programming Helps Female Computer Science Students. *ACM Journal on Educational Resources in Computing*, 4(1), pp. 4–es. doi:10.1145/1060071.1060075

Wigfield, A., Eccles, J. S. & Rodriguez, D. (1998) 'Chapter 3: The Development of Children's Motivation in School Contexts', *Review of Research in Education*, 23(1), pp. 73–118. doi: 10.3102/0091732X023001073.

YouTube. #LikeAGirl. https://www.youtube.com/watch?v=XjJQBjWYDTs [Online, Accessed 21 February 2022]

Design and project approaches in computing education

Mick Chesterman

Introduction

The creative processes involved in undertaking computing projects in an educational setting have a significant potential to deliver transformative learning experiences. Pedagogies and frameworks to support design approaches in education are valuable for teachers to guide and maximise the experience of their learners. This chapter describes and explores some of the strategies that can be used to support the delivery of design- and project-based approaches. To start, I focus on the value of creative communities in the design process. I then outline three teaching techniques aligned with design-based approaches to computing projects. The second half of the chapter takes a broad look at some of the benefits and processes of project-based learning (PBL) and ends with some tactics for overcoming limitations in delivering PBL in the classroom.

The power of communities

A project-based approach to learning coding and computing can be supported outside the boundaries of formal learning. For example, enthusiastic older family members may take young people to Maker Faire or engage in other community coding activities. Family members may buy creative computing kits or access resources such as YouTube videos or online forums for specialist interests like robotics, games or other forms of digital making. However, access to this kind of computer enthusiast community is not available to all young people. The following initiatives aim to address this by providing entry points to community-oriented coding activities.

Code Clubs are out-of-hours school clubs run by teachers or volunteers. Running a code club is a good way to build a lunchtime or after school community of coding enthusiasts. Code Club was originally an independent organisation but is now part of the Raspberry Pi Foundation. A large number of high quality, colourful and attractive resources are supplied free of charge on their website.[1] The resources provided can be printed out and serve as step-by-step tutorials. Resources also contain challenges that encourage further experimentation by learners.

Coder Dojos are monthly events run by volunteers at weekends. They often focus on creative, engaging computing. I have volunteered at some events and been impressed by the dedication and inventiveness of the other volunteers. There is an inspiring diversity of activities at events. At a typical Coder Dojo, some learners will use existing resources to support multimedia coding in Scratch, while others will try out new and experimental work, perhaps hacking Minecraft, creating games with code engines or trying new physical computing projects. Often volunteers bring their own children, who act as guinea pigs and test activities before the volunteer brings them to a Coder Dojo. To find out more, visit the Coder Dojo website.[2]

The Coolest Project is another project of the Raspberry Pi Foundation. It takes the form of an online competition and a real-world showcase that individuals or teams apply to take part in. The Coolest Project helps generate a framework and motivation for projects. This experience allows students to tackle problems in a radically different way than typical classroom teaching. Projects provide opportunities to engage in authentic coding practices, including designing for real users, collaboration with other students, project planning, debugging faulty code and repeated revisions to fine-tune the desired result.[3]

Coder Dojos are family-focussed events that take place outside of school and are therefore less easy for teachers to engage with. They are, however, a good source of inspiration for teachers looking for creative project ideas for their classroom. In contrast, both Code Clubs and the Coolest Project are suitable for computing teachers to run inside schools. Running lunchtime or after-school projects, while not reaching all pupils, can be a great way to showcase the engaging and creative nature of hobbyist computing projects.

Communities in educational theory

The power of communities has been highlighted by academics as part of what is known as the social turn in education. This is a turn away from more individualised ways of learning that focus on the transfer of knowledge from the teacher to the pupil. Instead, the focus is on how learning happens through participation in communities and culture, or in other words a socio-cultural approach. Community in this educational context can motivate and provide support for participation in a creative process. Barbara Rogoff (1994), a researcher of socio-cultural approaches to education, has described an educational process she calls *Communities of Learners*. Rogoff sees this approach as radically different from both instruction-based models of learning and pure discovery learning (where learners are left to their own devices). In communities of learners participants have different levels of expertise and varied roles in a learning system working towards an authentic goal. Rogoff notes that observing this kind of learning can be confusing to teachers and parents used to more instruction-based approaches. Such a learning community in full swing can seem chaotic. However, complex and productive learning is happening in ways that we, as teachers, may be unused to. This chapter helps unpick some of these practices and explore ways that educators have structured their learning environments to take advantage of this powerful approach.

> **Checklist - Applying a Community Approach in the Classroom**
>
> Are you making the most of the power of communities in your classroom? Before you start your next unit of work, ask yourself some of the following questions.
>
> - Are there regular opportunities for learners to work together during your unit of work? How often will students give and receive peer feedback?
> - Are there examples of similar work from other students available for your students to examine and perhaps build upon?
> - Can you draw on the roles or identities that students have adopted in their previous school or home activities? Are learners able to reflect on the specifics of those roles to contribute to the effectiveness of their engagement in teamwork?
> - Can you help your learners make connections between their computing activities and other professional or enthusiast communities outside of the classroom?

Design-based approaches in computing education

Designing as a discipline involves a community of both producers and users. Design-based approaches have been widely adopted in software production, the creative industries and wider business contexts. These design principles and practices are also relevant to education. For example, we may be used to seeing students motivated by producing something for a real audience. Design projects allow students to develop important 21st century skills like problem solving and communication, and creatively responding to real life contexts. In the following sections I will explore the design-based approaches of iteration, design patterns and the Use-modify-create (UMC) model.

Teaching technique – iterative design

Iterative design involves repeatedly coming back to reflect on the initial outcomes of creative goals and revising them based on results. The process involves goal setting, creating quick prototypes, user testing and evaluation, revision and ongoing reflection. The process is iterative in that testing and revision of the prototype design can be repeated until the desired result is achieved. Iteration is also a key part of a more general scientific method of testing a hypothesis and revision based on your analysis of the results. The idea of a repeated (iterative) spiral approach that both deepens understanding and improves the end results is popular both in

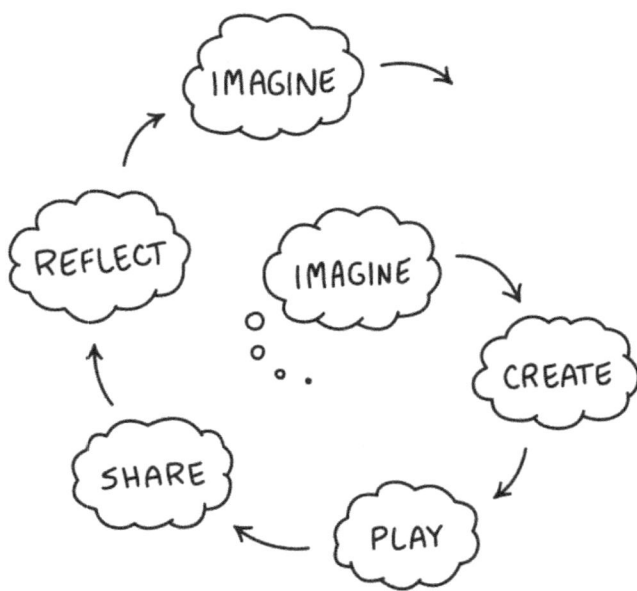

Figure 8.1 Diagram of creative learning spiral: Imagine – Create – Play – Share – Reflect – Imagine

Source: This is with permission from Resnick (2007).

education and industry. In software and design industries, it is often referred to as design thinking and agile approaches. In education, this approach is referred to in concepts like the spiral curriculum and the promotion of student mastery (Figure 8.1).

Resnick (2012) describes the foundations of the design-based approaches in education as: engaging in design activities, exploring personally meaningful topics, collaborating with others and deepening understanding through reflection. The key reason to adopt these principles is to increase engagement via sustained participation in computing projects for a broad range of learners.[4] One of the sources for sustained engagement is when, as part of the iterative process, learners are able to test and then revise their creation or experiment based on their own evaluation. Another factor is the importance of a community in the design process, as a real audience for creations, as a source of inspiration and as peer evaluators in the testing process. The above principles embody key elements of inclusive practices contained in universal design for learning (UDL – explained in more detail in Chapter 3), including allowing students to demonstrate their knowledge in a multitude of ways and allowing students to follow their own interests and motivations (Capp, 2017).

Teaching technique – worked examples and design patterns

Design patterns are most commonly used by computing students in higher education to teach object-oriented computing, but they are also useful for all levels of learners. Design patterns are rooted in real-life incidences of problems that are often solved in a particular way. They are concrete examples of coding principles in context. Design patterns can help the development of coding communities if more experienced coders take the time to document the patterns they use in an accessible way for novice coders. In computing education there are similarities between design patterns and a technique called *worked examples*. The National Centre for Computing Education (NCCE) has promoted working examples as a classroom activity by creating a Quick Read document for teachers on the subject.[5] Both worked examples and design patterns act as a way to demonstrate underlying principles in practice. For both approaches, showing working code used in a particular context helps students to analyse what makes it work and why it is a suitable solution.

For educators the use of design patterns and worked examples can help support learners develop coding proficiency by providing scaffolding and through modelling good design decisions. However, one of the challenges for teachers of using worked examples and design patterns is how to integrate them into student-led design challenges. You may be able to create a menu of printed or online patterns or examples that students can draw on as needed. Common examples can be modelled to the whole class when it is clear that many students will benefit from that approach.

Teaching technique – the use-modify-create model

The teaching technique UMC has the potential to both limit learner anxiety for novice coders and to scaffold the acquisition of coding and computational thinking concepts (Lee et al., 2011). A breakdown of each stage follows:

Use: In the *Use* stage, coders build familiarity with coding interfaces, code structures and syntax through scaffolded approaches that involve interacting with the program code and what it produces.

Modify: In the *Modify* stage, learners progress to working on real projects created by others. Learners deepen their knowledge of coding structures and practices by altering existing projects and templates to suit their own aims.

Create: After novice coders become more familiar with patterns of code design in use in the modify stage, they can progress to replicate such patterns in other code that they create from scratch.

A study involving five hundred 9- to 14-year-olds found that the UMC approach can balance a structured approach with more student-led exploration (Franklin et al., 2020). The researchers also found that the students enjoyed the UMC approach as they had more choice and agency in the process. Similarly, other research compared UMC with a starting-from-scratch approach and found higher student engagement for those in the UMC group (Lytle et al., 2019). The researchers found that because students using UMC had more time to play around with code, they were able to add their own personal touches and this ownership over the code sustained their continued engagement.

Researchers Kafai and Burke (2013) argue that a shift from writing programs from scratch to modifying and remixing them is in line with socio-cultural teaching approaches. They coin the term computational participation to reflect this change of focus. They also note that such remixing is helped by online coding communities, especially those aimed at novice coders like the Scratch community (described below). They encourage educators to avoid focussing solely on technical possibilities of coding environments to embrace the potential of online coding communities despite associated challenges. One challenge of teachers embracing remixing practices is to distinguish the legitimate remixing of work from less productive kinds of copying. In addition participation in online communities requires learners to concurrently build both technical and participatory skills which may also be challenging. The following case study examines ways in which the online Scratch community is facilitating design-based learning. It makes a strong case for the value of these approaches and asks how some of these benefits can be made more inclusive for learners who would struggle to take part in such a community independently.

Case study – how the online scratch community supports design-based learning

Scratch is educational software that uses a block-based coding approach and a set of tools to develop audio and graphical assets to help in the creation of multimedia coding projects.[6] Scratch has an extensive community with over 75 million users who have created 80 million projects. Activity increased during COVID restrictions in 2020 and 2021, with over 20 million user comments in the month of March 2021 alone. The online community allows young creators to connect with others to share and get feedback on their work. Such community interaction sustains repeated effort to build student mastery in the form of fluency in the design and coding processes. Here are some of the key features of the online Scratch community, along with tips to integrate them into your computing teaching.

High diversity of creations: The process of keeping such a large community up and running and safe for young people requires a lot of resources. However, the effort is justified, as it has become an extremely rich source of inspiration for young creators. A simple search of the site for projects like games, creative greeting cards, storytelling projects and pretty much any digital product you can imagine will yield a multitude of results. As teachers we can draw on this resource to demonstrate diverse creations and encourage our learners to adapt existing work based on their own interests.

Diverse ways to participate: Learners can engage with the online community in a great variety of ways. Learners may just play or comment on the games of others. They may use it to create their own projects but not engage in the more social elements of the creative process. They may, like a smaller section of the community, become extremely active in creating and collaborating with others on shared projects.

Encouraging project iteration: Scratch encourages the remixing of others' projects and makes it easy to create different versions of your own projects. This encourages sharing drafts for feedback via peer comments. This public sharing and feedback have been shown to encourage the development of new features in users' creations. In the classroom, teachers may need to balance the more disruptive possibilities of the ability to publish inappropriate material online with the potential to build student autonomy and reflection.

Supportive and authentic audience of fellow creators: Due to the high numbers involved in the online Scratch community, there is a good chance of finding peers who are also interested in specific subject matter. Peer collaboration between community users is motivated by these shared interests. The potential and depth of collaboration in this community can be impressive. Roque and colleagues (2016) have described this in detail.[7] The researchers describe how individuals find each other and group together by forming *studios* and then recruit other members to work on joint projects. This is sophisticated behaviour, which mimics real production processes. It is carried out by young people with a high degree of independence. However, the researchers concede that such collaborative production is only carried out by a very small proportion of the online creators. Teachers should be aware of a key challenge identified by the researchers, namely, how to replicate the benefits of collaborative community activity for young people who have less experience or confidence. In response, Roque (2016) went on to develop related programmes including online project exhibitions, competitions and off-line family-based projects to engage under-represented groups. As educators, we can take inspiration from this process of replicating the highly engaged, organic feedback and support of the chaotic online community into a more offline and structured design-based environment. The second half of this chapter addresses ways you may be able to rise to this challenge using project-based approaches.

Reflection point – using design-based approaches in the classroom

You can ask yourself the following questions to try to help you use some of the beneficial aspects of design-based approaches in your classroom.

- Are learners able to explore exemplar materials to inspire and shape their creative ideas?

- As they plan, are learners able to think about and articulate the perspective of the real or imagined users of the designed projects?

- Are learners helped to come up with ideas through ideation techniques that scaffold the creative process?

Follow-up resources: I have created several online courses that explore hands-on ways to use design thinking in education and community work as part of Manchester Met's Rise initiative, which provides students with opportunities to gain new skills.[8]

Projects and project-based learning (PBL)

PBL is a wide set of approaches that seek to facilitate learning through undertaking practical projects. Students often complete projects in groups. Students develop knowledge and skills in the context of a real or simulated problem that they must solve. Using projects to support learning is one of the 12 teaching principles advocated by NCCE.[9] In the next section, I cover the characteristics and potential of PBL.

Researchers Blumenfeld and colleagues (1991) argue that school disengagement is caused by work that bores students. They found that project work incorporating learner choice and involving real outputs is motivating and can sustain student engagement. The pervasiveness of digital products in our lives offers a wide range of choices for computing projects, including websites, games, wearable technology, phone apps, robotics and other physical computing projects. Thus, computing education is an excellent vehicle for a project-based approach to learning. However, Blumenfeld's research cautions that implementing PBL in classrooms is not straightforward. I cover barriers to PBL and ways to overcome them in the final part of this chapter.

Academics have worked with expert practitioners to create PBL frameworks to help teachers plan and deliver projects and recognise the complexity of some of the learning that takes place. The following outline of PBL elements is a synthesis of several of these frameworks with additional commentary on how this may apply to computing projects.

Challenge: The focus of the project should be a relatable problem or question that does not have one straightforward solution. Software and electronics projects that fit this brief are thus very suitable candidates.

Authenticity: The real-life relevance of projects helps engage students as they make connections to their interests and communities. As mentioned in the section above, many forms of coding projects, including phone apps, websites and games, have real and relatable goals.

Sustained and collaborative work: Adequate time must be allocated for sustained project work. Students should work together and be given the chance to revise their projects. This is perhaps one of the greatest challenges to delivering computing projects in a school setting.

Public project: The creation of a shareable, public object helps learners focus and design for others. It can also act as a focus for discussion within the classroom. Sharing computing projects within the classroom could be supplemented by presentations to a wider audience.

Student voice and choice: Giving student's choice over the focus of their project increases their engagement. Participation in discussions about project direction builds student autonomy. The high level of saturation of digital products into the experience of many young people's day-to-day lives can help shape students' interests.

Reflection and critique: Self-reflection may be informal at times but also guided by classroom processes like learning journals. It can also involve peer feedback or input beyond the classroom to bring authentic perspectives. Reflection could also happen in digital form via an online journal or templated digital documents.

Teaching technique: Supporting PBL in computing classrooms

PBL and other design approaches are aligned with UDL in many ways. For example, they all benefit from a collaborative community and a real or imagined audience for public products as a way to increase learner motivation. They also advocate for structures for students to chart their progress and support feedback. One of the central tenets of these approaches, the importance of learner choice in projects, is also one of the most challenging for teachers to enact. One critique of PBL, especially where it involves student experimentation and student discovery, is that it can be chaotic and that it is therefore challenging to communicate high-level concepts. PBL also requires skills, support and planning that are very different from traditional teaching and may therefore be difficult for teaching staff to implement. For example, practitioners must build their ability to switch between facilitating students operating freely to then guiding them in the process of revision and critique. Having resources and clear stages in your project plan to help this process is vital. This section outlines the typical stages of PBL and how to adapt them to a computing context.

Start with a driving question or mission: The project goal for computing projects is often to create a digital product in response to a need or design brief for a specific audience. Add detail and link to real-world problems at this stage to maximise learner engagement. Decide the limits for students projects and outline these clearly from the start to avoid having to dampen their enthusiasm. For example, if creating a 2D game instead of a 3D one is better suited to the technical abilities of students, then be clear about that limit from the start of the project.

Designing a plan and resources for the project: Decide what part of the curriculum the project work will develop. Use your knowledge of the curriculum to put resources in place to support the learners as they undertake the project. Not everything needs to be explicitly taught if you can signpost your learners to those resources. Having an online or paper-based resource bank for students to access and navigate can be extremely useful.

Monitor pupil progress: As the project unfolds, keep students on track by having a realistic schedule for project stages. Check that you are consistently signposting students to the relevant resources for the project choices they have made and for the stage that they are currently undertaking.

Assess emerging project processes and outcomes: Ongoing feedback and assessment are vital. Build in opportunities for reflection, peer feedback and revision. Can students share prototypes of their digital products? Can you support them to recognise if they are working effectively as a team? How can you support them in making connections to the underlying curriculum knowledge?

Evaluation: You may evaluate the end piece of work created by students, the way they have worked together and the skills used to undertake different stages of the project. You can validate what the students have learned and identify areas for future development.

If you would like detailed information and case studies on PBL, there are online resources provided by numerous organisations, including Edutopia,[10] PBLWorks[11] and the UK-based Edge Foundation.[12]

Creatively overcoming limitations to PBL

This section considers barriers to PBL and provides tips and strategies that have been used by other educators and researchers to overcome them in the context of computing.

Sustaining the effort – Time challenges: In research on barriers to undertaking projects in schools, teachers commonly cite time restrictions due to curriculum pressures. Resnick and Rusk (2020) suggest that, if possible, double lessons are helpful for hands-on coding work and to allow the design process time to unfold. They also advocate that at times a whole term should be devoted to undertaking a project. This sustained effort allows pupils to return to tweak and improve trickier coding and design challenges, thus supporting an iterative approach. In addition, cross-curricular projects may free up more time by linking with other subjects that are allocated more time, especially in KS3. For example, you could link a computing project with mathematics by asking students to create a game that teaches mathematical concepts. This could both deepen students' learning of particular mathematical concepts and allow for the kind of repeated hands-on practice that builds coding fluency.

Advocating the value of PBL for inclusion: As with design-based approaches, PBL aligns well with UDL, specifically in the way students can bring their own interests into their creation of a public project. Both frameworks encourage teachers to guide and support project work with scaffolding and ongoing feedback. As educators, we can highlight the importance of creating inclusive classroom environments to our line managers and advocate for the time, training and resources needed to undertake project-based learning.

Artefact-based assessment: The tension between rote-learning approaches that may be used to prepare students to reproduce knowledge in written exam questions and the need for more fluid programming experiences raise an important question. How can some of the more flexible techniques for observing and assessing learner's progress be brought into classroom practice? As a possible response, the NCCE promotes the use of artefact-based questions (ABQs) to assess project work. ABQs are questions based on the digital or physical artefacts that students create as part of their projects. ABQs allow teachers and students to link the problems students have encountered and solved in their computing projects to the requirements of the computing curriculum. Teachers can focus on specific areas of the students' work and ask about details of the code structure and implementation. ABQs can also address design issues and processes. For example, how the project outcomes compare to the original goals, how feedback was implemented, group work and overcoming challenges and the design challenges.

Authenticity of projects: As mentioned before, computing is blessed with the potential to create digital and physical projects that are recognised by or relevant to young people. However, sometimes the process of finding authentic audiences and processes to motivate learners within a school setting is not simple. The following activity may help bring authenticity to student projects.

Activity – Meeting the Challenge of Authenticity in the Classroom

Here are some tactics you may be able to use to link projects to real issues and activities beyond the classroom.

- Draw on community members to set a local challenge that resonates with your learners.
- Use other members of staff in other subject areas to pose a school-based problem. This could be subject-specific or a pastoral or cross-curricular issue.
- Establish links with industry or social enterprises to set an authentic challenge within a work context.

If you draw on experts, staff or community members, they do not need to be there for the full term of the project. You can use visits or video calls at the start and end of the project. Most importantly, be sure to draw on the experience of students and use their ideas to shape possible responses to the challenge in the early stages.

Conclusion

Motivation, participation and peer learning do not happen in a vacuum. In this chapter, we have explored the value of creating a learning community of coders working on projects that are both authentic and linked to their own interests. To help this happen, we can draw on some of the rich research and resources available from different streams of practice, including PBL, UDL and design-based approaches. What many design and project-based approaches have in common is their focus on learner choice, sustained hands-on making and frameworks for facilitation, observation and assessment. For an accessible and convincing summary of project-based approaches and their adoption in a classroom setting see the review by Barron and Darling-Hammond (2008).

We have explored the tension between creative processes involving learner choice and teaching the more prescriptive requirements of the computing curriculum. To help bridge this gap the NCCE have created resources drawing on socio-cultural research to offer guidance on PBL, observation and pair programming. As teachers seeking to close an achievement gap between higher and lower achieving students, we have a responsibility to integrate such methods and evaluate their potential for our learners. I hope that this chapter has encouraged you to keep exploring authentic coding practices in the classroom and to share your experiences with others. To continue this journey, there are many forums where teachers share practice; these include CAS forums, blogs, Twitter posts and so on. In particular, there is a Twitter hashtag called #casinclude that encourages computing teachers to share inclusive practices. To fully explore the potential of projects, let's share how we have used design and PBL approaches in our work and encourage others to do the same.

Notes

1. https://coderdojo.com/
2. https://projects.raspberrypi.org/en/codeclub
3. https://online.coolestprojects.org/
4. https://mres.medium.com/ten-tips-for-cultivating-creativity-fe79e7ebb83e
5. https://blog.teachcomputing.org/using-worked-examples-to-support-novice-learners/
6. Scratch is available as a free download at https://scratch.mit.edu/
7. http://tiny.cc/scratch-community
8. https://rise.mmu.ac.uk/category/enterprise/design-thinking/.
9. https://teachcomputing.org/pedagogy
10. https://www.edutopia.org/project-based-learning
11. https://my.pblworks.org/
12. https://www.edge.co.uk/edge-future-learning/project-based-learning/

References

Barron, B., Darling-Hammond, L., 2008. Teaching for meaningful learning: A review of research on inquiry-based and cooperative learning. Book excerpt. George Lucas Educational Foundation.

Blumenfeld, P.C., Soloway, E., Marx, R.W., Krajcik, J.S., Guzdial, M., Palincsar, A., 1991. Motivating project-based learning: sustaining the doing, supporting the learning. *Educational Psychologist* 26, 369–398. https://doi.org/10.1080/00461520.1991.9653139

Capp, M.J., 2017. The effectiveness of universal design for learning: a meta-analysis of literature between 2013 and 2016. *International Journal of Inclusive Education* 21, 791–807. https://doi.org/10.1080/13603116.2017.1325074

Franklin, D., Coenraad, M., Palmer, J., Eatinger, D., Zipp, A., Anaya, M., White, M., Pham, H., Gökdemir, O., Weintrop, D., 2020. An Analysis of Use-Modify-Create Pedagogical Approach's Success in Balancing Structure and Student Agency, in: Proceedings of the 2020 ACM Conference on International Computing Education Research. Presented at the ICER '20: International Computing Education Research Conference, ACM, Virtual Event New Zealand, pp. 14–24. https://doi.org/10.1145/3372782.3406256

Kafai, Y., Burke, Q., 2013. The Social Turn in K-12 Programming: Moving from Computational Thinking to Computational Participation, in: Proceedings of the 44th ACM Technical Symposium on Computer Science Education. https://doi.org/10.1145/2445196.2445373

Lee, I., Martin, F., Denner, J., Coulter, B., Allan, W., Erickson, J., Malyn-Smith, J., Werner, L., 2011. Computational thinking for youth in practice. *ACM Inroads* 2, 32–37. https://doi.org/10.1145/1929887.1929902

Lytle, N., Cateté, V., Boulden, D., Dong, Y., Houchins, J., Milliken, A., Isvik, A., Bounajim, D., Wiebe, E., Barnes, T., 2019. Use, Modify, Create: Comparing Computational Thinking Lesson Progressions for STEM Classes, in: Proceedings of the 2019 ACM Conference on Innovation and Technology in Computer Science Education. Presented at the ITiCSE '19: Innovation and Technology in Computer Science Education, ACM, Aberdeen Scotland UK, pp. 395–401. https://doi.org/10.1145/3304221.3319786

Resnick, M., 2007, June. All I Really Need to Know (About Creative Thinking) I Learned (by Studying How Children Learn) in Kindergarten, in Proceedings of the 6th ACM SIGCHI Conference on Creativity & Cognition, pp. 1–6. https://dl.acm.org/doi/abs/10.1145/1254960.1254961?casa_token=22n5lCY9h5IAAAAA:CLDuH44zSkZD0UJjIEHTYZG8L-7RpBKkPosIX4Tzg9QvGluv_rcXxIf6yBQ4aaZqN9r8K3d0YzzV_

Resnick, M., 2012. *ScratchEd: Working with Teachers to Develop Design-Based Approaches to the Cultivation Of Computational Thinking.* MIT. https://web.media.mit.edu/~mres/proposals/NSF-ScratchEd.pdf

Resnick, M., Rusk, N., 2020. Coding at a crossroads. *Communication of the ACM* 63, 120–127. https://doi.org/10.1145/3375546

Rogoff, B., 1994. Developing understanding of the idea of communities of learners. *Mind, Culture, and Activity* 1, 209–229. https://doi.org/10.1080/10749039409524673

Roque, R.V., 2016. Family creative learning : designing structures to engage kids and parents as computational creators (Thesis). Massachusetts Institute of Technology.

Roque, R., Rusk, N., Resnick, M., 2016. Supporting Diverse and Creative Collaboration in the Scratch Online Community, in: Cress, U., Moskaliuk, J., Jeong, H. (Eds.), *Mass Collaboration and Education.* Springer International Publishing, Cham, pp. 241–256. https://doi.org/10.1007/978-3-319-13536-6_12

Industry perspectives

Louise Hayes and Eleanor Overland

Introduction

When thinking about the curriculum in computing education, it is important to consider the next steps for the students. The best computing teachers are those who are up-to-date with developments in the computing industry and understand how computing knowledge is beneficial to all aspects of life. Those teachers are best able to develop a meaningful and ambitious curriculum and provide appropriate information and guidance on further education and careers in computing.

This chapter explores perspectives from industry and the research community. The contributors are from across England and provide some useful snapshots of developments at the time of writing. Whilst these contributions are valuable they only provide a small piece of the jigsaw. Computing changes considerably over time, and needs and skill shortages can vary between regions. The later part of the chapter provides some advice as to how teachers can make links themselves with local industry and researchers to gain their own industry perspectives. Working on a local level can increase the relevance of the curriculum and also provide opportunities for specialist input, enrichment and work related learning.

In this chapter, guest contributors were interviewed and have kindly shared their thoughts on developing computing education and ultimately broadening the computing workforce and making the industry more inclusive. Their own expertise is varied and includes experience as employers, researchers and industry-based trainers within computing. The questions used to shape the interviews were all similar in nature to start with, but conversations developed in interesting ways to include individual experiences and viewpoints.

Interview 1

Katharine Childs, Raspberry Pi Foundation

This interview was held with Katharine, Programme Coordinator at the Raspberry Pi Foundation, where she leads the coordination of the research programme called Gender Balance in Computing.

Gender imbalance in computing

Katharine reported that the subject of computing is still one of the subjects where women and girls are least well represented. Compared to the other sciences, such as biology or chemistry, where there has been a bit of a shift, and even to some extent mathematics; computer science really does stand out with the imbalance. This aligns with reported figures on the gender balance from The Wise and Roehampton reports. As mentioned in earlier chapters, Katharine commented that the figures are pushing up towards the high teens, with 20% of Computer Science General Certificate of Education (GCSE) pupils now being girls and slightly lower at A level.

Underlying attitudes and beliefs

Katharine stated that it's around students feeling and having attitudes and beliefs that they can succeed in a subject. It's not about whether girls or women have the cognitive aptitudes to succeed in computing and computer science, because they undoubtedly do. It's about the underlying attitudes and beliefs that lead them to feel like computing isn't a subject for them currently and what can be done to change the subject so that it is more inclusive.

Computing is an abstract subject

As a subject, computing contains lots of abstract topics, especially as you go further up the curriculum into GCSE and A-level computer science and beyond. Research is starting to suggest that if you can link computing to solving real-world problems, then it's more likely to be a subject that sparks an interest in a more diverse group of pupils.

What makes an inclusive real-world problem?

Inclusive real-world problems that spark interest in a more diverse group of pupils are themes emerging from recent research. One such paper is Hidi and Renninger (2006) four-phase model of interest development. This is a model, Katharine states, that

> moves from a short-term situational interest where there's something about the way a subject is presented that sparks some initial interest, and then over

time and with the right learning environment, that moves to a more embedded, longer-term interest where a pupil then picks up and goes right. This is something that I'm really interested in there. I could see a use for this. I can see a purpose for studying this subject.

Katharine recalls studying computer science and looks back at the sorts of problems that she used to solve, such as scheduling aircraft or looking at a banking system, which are actually real-world applications of computing. But the problem was that they're still so far removed from everyday life,

> That they don't necessarily have enough applicable context that related to the types of work I was interested in pursuing. So yes, I could see that if I wanted to go and be a computer programmer in the bank, then I could carry on this course, but that wasn't necessarily my career goal or my aspiration.

The idea of real-world computing focusing on solving problems in local communities is a significant shift in the classroom and, as she says "really kind of innovative". Here, the mention of the *Technovation* program is an example of that very thing, which was particularly aimed at women and girls thinking about their own communities and developing apps.

Youth culture and local communities

In earlier chapters, we have talked about using technology safely in classrooms. What is interesting here for computing educators is, as Katharine says,

> the idea that teachers have to be fully up to date with youth culture is a bit of a misconception. This isn't about making computing trendy. This isn't about suddenly getting on TikTok and doing whatever the latest craze is to meet young people where they're at. What it's about is thinking about the pupils that you've got in your classroom and allowing the classroom to be a safe space where they feel that they can come as individuals with their own interests, with their own cultures.

Adapting this to your own lessons is about getting to know your students and who you are teaching. As a resident of Derbyshire, this was where Katharine would look to link to her local community when she was a primary computing teacher. Also of interest is the *Roots project*, which at the time of writing this book, is a new research project developed by the Raspberry Pi Foundation that explores how computing can be made culturally relevant to pupils.

Katherine identified that we are moving into a new era of learning brought about and even accelerated by COVID and the use of technology in the classroom. Katherine concluded the interview by identifying the current time as a real opportunity for teachers to increase diversity in computing. But there is also an acknowledgement that there is still a lot of work to do in this area, particularly in supporting teachers with the knowledge and confidence to support change.

Interview 2

Dale Lane, author of machine learning for kids and employee of major international technology company

Q.1: What are the key digital skills that you think young people should be leaving school with?

I have a slightly warped view of the world here because of the kind of projects that I work on, but where we see the biggest skills gap when I get involved in recruiting and where we still struggle is around processing and handling data.

We have the technology to do the projects we need, but all of these projects need data that is sliced and diced into the right shape, and we need people who can do that. We must also understand that the needs and ideas of people, our customers, are super valuable.

People who can do powerful work with spreadsheets have the right kind of skill base. Businesses, both commercial and public, are looking to make data-driven decisions. That all stems from being able to know where to go and find the data. What do you have to do to clean up the data? Understanding how you treat data, how you process data, and the risk of bias in a data set. It is important to know that you don't just go to the first thing you find on Google and base your entire project around that. The risks that come from doing that including the risk of bias. There have been so many mistakes where we've based projects on datasets that have been collected on a particular type of person or demographic group. Whether it's a particular gender or social class or whatever, and then built systems that either discriminate against or exclude chunks of the population. Often, this is not done maliciously; it's because of a lack of awareness. It's because people don't understand the implications of where they get the data from, how they process it and how they treat it.

Q.2: Do you think more schools could be doing more to prepare children for AI in their lives?

There are business people around the world who don't think of themselves as developers, but who are writing macros, add-ons and all sorts of formulas in spreadsheets and what they are doing is processing data. They just don't think of it like that because they think "I'm not a developer".

A number of tools keep coming up around the AI and machine learning space that try to use spreadsheet metaphors to apply machine learning algorithms because that's held up as the "how to get people to do programming without them realising they're doing programming" approach. But the objective is, we're going to give you a massive chunk of messy, incomplete data or we're going to give you two separate data sets. Your job is to correlate them against each other, clean it up, weed out the noise and find the missing data. Whether you do that with the tool or you're writing code or scripts to do it, it's the capability and understanding that are needed, not necessarily coding skills.

There is definitely a huge need for data skills, and, often, what we're finding is that we're having to train people to do that when we recruit them because there aren't enough people with those skills. We spend a lot of time and effort developing what we would call "no code tools" or "low code tools" as a tool that is accessible to business users and wraps programming up in some friendly interface.

The other concern comes back to ethics. The managers, the decision-makers and the policymakers need a good understanding of ethics. It's about them knowing where the opportunities are and what the risks are. What we do see when we start getting involved with companies in projects is that the only people around the table who understand what machine learning are the other developers. The people who are leading the projects don't know what questions they should be asking. They are not able to think about the safeguards they should put in place at the start of the projects. This comes back to the need for basic literacy about technology. So it's not just the developers who are building the systems who understand the implications, but the people who are commissioning, funding and approving these projects who need to know.

Q.3: Is there anything more than you want schools to be doing to prepare young people for the AI careers workplace?

If you want to be a project manager or if you want to be in management in any kind of industry then digital literacy is just as important as any other literacy.

As part of our recruitment of a project manager, there isn't a project that doesn't involve an element of data science, machine learning and artificial intelligence (AI) understanding. The safeguarding side of that around ethics and risks is part of the training for every project manager.

I think it was a few years ago when the story came out; Amazon employees were listening to recordings from Alexa devices to improve their training, and everyone who works in machine learning said, "of course they were, that's how every machine learning project works". How else do you improve the training? When the story emerged in the media, it raised lots of questions about data security and the ethical use of data. For most developers, they would put out an early version of their machine learning system and then try to keep track of where it gets things wrong. They capture all of the examples that went wrong and add those to the training so it gets better at doing that sort of thing in the future. So for Alexa, of course Amazon would have collected all the recordings of times where Alexa gave the wrong answer or the user became annoyed. They would add that into the training so that next time it would provide a better answer. That was just obvious, but judging from the massive backlash in the press, it wasn't obvious to 'normal people'. Even if children aren't the developers that are going to build such a system, when they buy this kind of device and put it in their home, they should understand what is happening.

There are far more important ways that AI will affect our lives. Whether it's increasingly being used in recruitment and evaluating mortgage applications or all sorts of other sectors, if these systems are affecting our lives, we should be explaining to kids how the world around them works and what these systems are doing.

It's right on the cusp of being totally easy for anyone to start developing a simple AI system, so it's hard to identify what kids need to learn today. I do think we're within a small number of years of having a spreadsheet-like tool that does machine learning. That is something that anyone can use that doesn't involve paying a company to come in and do something complicated for you. So I'm fumbling because, well, how do you prepare kids to use a tool that hasn't been invented yet, but I'm convinced it is right around the corner.

Part of the motivation for the education resources that I've built is getting kids to get hands-on with the technology. I've wrapped it up in Scratch because that's what they seem familiar with (see Chapter 2).

For secondary-age children you could want to go more towards using Python or other text-based languages, but I'm hesitant about whether all secondary school students need to train their own machine learning model in Python. It's kind of cool to code, but do they need to? Probably not, but they do need to understand how the programs work and the implications of them being done badly. It is important for them to get to see it for themselves, to build something for themselves and to train a biased model. They understand how bias happens, that sort of thing, rather than just telling them about it.

Dale ended the interview answering the question of how the appeal of technology can improve the sector, where he stated

> The people who are working within the technology sector don't just look like me, but this can be overlooked at the development stage. Face recognition, systems were working better with white middle-class men because lots of development is coming from the West Coast of America. It is a really narrow subset of the population that is building the systems, so they work well for that population group. But you still keep hearing about issues, like people with dark skin who put their hands under these automatic taps and it doesn't work. It's just a hundred little things like that, and the only way that's going to change is if we broaden the pool of people who are helping to build this technology. So that's part of why we do a lot of outreach work to try and broaden the appeal of technology as a sector. We think it will make the technology better, and we will end up with better systems as a result of having a diverse group of people build it.

Interview 3

Arlene Bulfin, Director of People Development, ANS Group

Tom Robinson, Head of Apprenticeships, ANS Group

ANS is a technology company with approximately 700 staff, based in the Northwest of England.

They work closely with schools to provide enrichment visits, work experience placements and learning experiences for teachers and trainee teachers. Their apprenticeship provision is graded outstanding by Ofsted, and ANS has a strong commitment to diversity and inclusion.

Digital skills for school leavers

We started the interview by asking about the digital skills that they thought young people should be leaving school with. Tom responded to the question saying that what they needed as a company was a wider skill set rather than just apprentices coming to them with digital skills. He believes that there are some skills that they are able to teach as part of the apprenticeship program, for example, how a network works. However, what was more problematic for them was the lack of soft skills among school leavers; as Tom says "I think it is a massive thing for us". Arlene supported this with the comment that as an organisation they were less worried about the tech, which she saw as "the easy part to teach", but saw communication skills as key. As an organisation they focus on building skills that ensure their apprentices can operate in a big forum, or articulate an idea in a meeting, or be able to share an idea they have with a group of people. This is a significant change in what is required in industry, and the ability to be able to explain a concept to an audience, that is both expert and non-expert in the technology field, has been identified as a need. Acknowledging that is not only a school leaver issue, as it is the same for students who come to them with a computer science degree, or similar.

A generation of consumers

> We show them Tom's brilliant diagram. In that it flashes and you can see all of the cable that runs around the world under the sea. It actually shows what puts us online.
>
> (Arlene)

This generation of consumers, Arlene sees "we are on our phones on our devices connecting and consuming, but it's the understanding of the concept of what is delivering the Internet, whether it's cloud skills or whether it's cloud concepts which would be so relevant now". Summarising some of the points that are made by Arlene and Tom, we can see that schools and the curriculum could include:

- A wider view, to know more about the hardware and how it all works together. Also, having an understanding of what a server and the concept of the internet actually are would be helpful; let's talk about everything to do with the server and link it in with all the other aspects.
- What happens when you pick up your phone or when you connect to a website? A server is someone's server room somewhere in the world.
- The curriculum breaks the subject up, and pupils are learning concepts in chunks – rather than the big picture. Know how a piece of RAM and the CPU all work together and move on to the next thing.

A missing piece in skills and understanding is around data. An understanding of storage, such as how many pictures and videos we take and store, where it all is, how it is managed and what impact that has on their lives going forward. The concept of the digital footprint online and having an understanding of the impact of that on their future. This was something that was key to Arlene's lessons whilst a school teacher herself when it used to be in the curriculum. A lesson would be how to lockdown your social media. How much time do you spend online? Arlene and Tom find that school leavers have amazing qualifications, but they are often shocked at the innocence in the use and lack of understanding of social media by their apprentices. Having said this, they have a deep understanding of how operating systems work are being able to get into the command line.

A consideration for computing teachers is their input into cross-curricular computing. Where are pupils discussing this in their lessons? If they take a photo, put it on Instagram, and delete the photo off their phone, where is the data for that photo? And who owns it now? These are skills that enable them to understand the bigger picture and are applicable in industry. They both accept that coding is clearly important, but having a wide reach of all those skills focuses more on the code inside.

Perhaps this is reflective of having an understanding of what is needed in industry (as seen in Chapter 8), which might avoid statements such as "I want to be a software developer", when they don't really know what that means, according to Tom.

The interview moved on to discuss other aspects of the curriculum. There may be a further discussion point around safeguarding. By opening up the conversation, rather than trying to block technology, we can educate children, and there will continue to be other platforms that they access in the future. Arlene recalls a homework task that she used to set that she was asked to stop doing by her

Head of Department. The objective of the lesson was to discuss privacy settings on platforms

> in the following week's lesson I would put pictures up when they walked in the door ... a family holiday or whatever 11 year olds are putting on Facebook at the time and they will come in and say. You can't do that. You can't. You can't take my picture and put it on a slide in this class.

This resulted in a shift in pupil thinking and an understanding of the need for privacy settings "there's massive learning there, isn't there? You're not happy for your teacher with no malice or malicious intent to look at your photos. You don't know who's looking at your photo" (Arlene).

Schools preparing young people for the workplace

Both Tom and Arlene (who themselves are have been teachers in Secondary Education in the UK) mentioned job descriptions, programming tunnel vision and applicants seeing coding as "cool" as a cause of imbalance in recruitment. Stripping back job descriptions has allowed them to widen their pool of applicants: "we never worked with project managers or scrum teams, and now that's something that we realised is going to absolutely drive the business forward". Arlene

Having coding as a large part of the curriculum is right, they think, as it is important to industry. AI, network security, firewalls and cloud services were also mentioned as being important. Summarising the points that they made, we can see that:

- Building teams with communications skills, being able to empathise with customers, and drawing out the problem that they want to solve allows them to support small businesses and pivot on projects if they need to.

- You can't be what you can't see, so they ensure that the school groups they work with have a representative group of pupils. Where are the female representatives in tech? Who are the inspirational people we are looking at? "How are girls age 15 going to look at a picture of Ada Lovelace? Think. Yeah. Cool. I wanna be her. Where's Elon Musk? The boys are; cool, I wanna be Elon Musk. So for me it's where are the Asian girls in tech doing amazing things? Where are the women and men in wheelchairs doing great things in tech?" Arlene.

- There is an opportunity for examination boards and publishers to look at how representative the GCSE textbooks are. How diverse and representative are the writers of the material?

We finished school with a better understanding of the pathways into technology as an alternative career path to being a teacher, doctor or lawyer. What do they actually look like? "Without the jargon and stuff, this is a cool, great job that you get" Tom.

As teachers, we are inspiring, encouraging and supporting our pupils in their future careers. Perhaps we often forget that our own career as teachers plays a part in this. We have a shortage of secondary teachers in this country. If we are to inspire the next generation of computing teachers, perhaps we have to look at why we went into teaching. Tom finishes the interview on this topic,

> That's why I went into teaching. If I'm honest, I'm not in a minority group. But I went into teaching 'cause it's a good job. I did a science degree, and that was my decision. I didn't have anyone else telling me. This is also a great path. And if that's not happening in other groups, then they certainly won't be thinking about those paths that are available to them.

A reflection point here for teacher educators as we want to help make those paths available to all.

Developing your own industry insights

Having read these interviews, you may well have started to reflect on your own practice. Does your curriculum resonate the views of experts in industry? Such reflections may result in some fundamental changes in curriculum design, including specific subject knowledge or the selection of qualifications. It may also lead to smaller changes, such as the context for classroom activities or the use of specific terminology.

In English secondary education, there is also a requirement for all pupils to receive appropriate career education information and guidance from the age of 11 (DfE 2015). Subject-specific support in this process can be really helpful for pupils and open up careers they may not have previously thought of. In developing a more inclusive computing education, an ultimate achievement would be to support diversity and inclusion for entry into the computing profession. Working with local employers and training providers can provide additional inspiration and support to ensure planning for next steps in computing is inclusive for all pupils.

A useful starting point to develop links with local expertise is to look at the previous destinations of your pupils and students. Local employers, apprenticeship providers and HE institutions will often be looking to increase the number and quality of applicants in the future. These links can often be made through the local careers service or even through your own alumni. Where destinations in computing are new for you, an alternative approach is to explore local businesses, especially those with vacancies for jobs or apprenticeships. Sometimes just the terminology of the job titles may be confusing for school or college leavers, so an initial conversation as to whether the roles are suitable for your cohort may be a useful starting point. This may then lead to further conversations about the required skills and experiences.

Also, be aware of some local and national organisations that may be able to put you in touch with organisations locally. Stem ambassadors, Computing at School (CAS), part of the British Computing Society (BCS) will all have local contacts. Most universities will also have outreach programmes to work with local schools in encouraging a more diverse and inclusive approach to university applications. Most organisations have a real commitment to making computing more inclusive and would really value links with computing teachers.

Reflection point – Revisiting your curriculum

The focus of all the chapters in this book is to raise awareness of the challenges of inclusion within computing education and empower teachers to explore and take actions to make computing more inclusive. Chapter 1 encouraged you to consider the intent of your curriculum and how it could potentially be made more inclusive. After reading the industry insights alongside the other chapters, it may be useful to revisit your curriculum design.

When reviewing your curriculum and approaches to teaching computing, key questions are pertinent to ensure inclusive approaches to computing education.

- Learning spaces.
- Content/qualifications.
- Keeping the curriculum relevant – applying established theory to new areas, e.g. AI and data handling.
- What does your data say?
- Is the terminology used within your curriculum appropriate? Is it inclusive in nature?
- Include underrepresented groups in resources/case studies.
- Pedagogical approaches for all types of learners, especially SEND.
- Readiness for next steps.

Teachers can make a difference.

Acknowledgements

A huge thank you to all of the contributors for their time and input into this chapter: Arlene Bulfin, Director of People Development, ANS Group; Katharine Childs, Programme Coordinator, The Raspberry Pi Foundation; Dale Lane, Father of two amazing kids, software developer for IBM in the UK, mobile and gadget obsessive, charity trustee and all-round geek; and Thomas Robinson, Head of Apprenticeships, ANS Group.

References

DfE (2015) Careers Guidance and access for education and training providers Careers guidance and access for education and training providers – GOV.UK (www.gov.uk)

Girls for a change, join the movement (2023) https://technovationchallenge.org

Hidi, S.E. & Ann Renninger, K. (2006) The Four-Phase Model of Interest Development, *Educational Psychologist*, 41:2,111–127, DOI: 10.1207/s15326985ep4102_4 https://methods.sagepub.com/base/download/BookChapter/a-practical-introduction-to-in-depth-interviewing/i531.xml

Kemp, P.E.J. & Berry, M.G. (2019) *The Roehampton Annual Computing Education Report from 2018*. London: University of Roehampton.

Lane, D. (2020) Machine Learning for Kids, Machine Learning for Kids: A Project-Based Introduction to Artificial Intelligence: A Playful Introduction to Artificial Intelligence. https://www.amazon.co.uk/Machine-Learning-Kids-Project-Based-Introduction/dp/1718500564

Sentence, S. (2022) The Roots project: Implementing culturally responsive computing teaching in schools in England. https://www.raspberrypi.org/blog/culturally-responsive-computing-teaching-schools-england-roots-research-project/

The Wise Campaign, Gender Balance in Computing (2022) https://www.wisecampaign.org.uk/what-we-do/wise-projects/gender-balance-in-computing/

Index

Note: **Bold** page numbers refer to tables; *italic* page numbers refer to figures and "n" indicates endnotes in the text.

Abid, A. 51
after-school projects 96
"After the Reboot" report 75
Agalianos, A. 20
AI betwixt 30–32
algorithmic abuse 47–50
algorithms 5, 45–53, 55
alternative perspectives on computing curriculum 12, 39, 40
annotated bibliography 73
Ann Renninger, K. 109
apps: data collection and analysis 63–64
artefact-based assessment 104–105
artefact-based questions (ABQs) 104, 105
artificial intelligence (AI) 1, 14, 19, 21, 25, 29, 45, 46, 53, 85, 93; in computing education 19–32; is inherently biased 45–55
Ashbee, R. 14
assessment data 60, 61, 80, 90
assessment for learning (AFL) strategies 65, 67
authenticity 105
authoritative voices: influence of 79–80

Banks Gatenby, A. 21
Barron, B. 105
biases 14, 50–55, 55n2, 111, 113; reproduction of 50–53

Big Data 47–50
blended learning 64–66
Bloodhart, B. 54
Blumenfeld, P.C. 102
Brennan, K. 39
British Educational Technology Exhibition (BETT) 5, 10
Burke, Q. 100

Chan, E. 25
chatbot 30, 51
ChatGPT 55n2
Cheryan, S. 88
classroom 1, 2, 4, 6, 19, 40, 66, 86, 89, 90, 97, 101, 102, 105, 110; design-based approaches 101–103; environments 2, 84, 87, 93, 104; inclusive and engaging 66–67
Code Clubs 95–96
code examples 40, 41
Coder Dojos 96
cognitive load theory 4
communities: in educational theory 96–97; power of 95–96
competencies for future 9, 43
computation: for unmaking concepts 21–28
computational concepts 39
computational doing 22
computational perspectives 39

computational practices 39
computational thinking (CT) 2, 34, 36–39; definitions of 39–40
computer science 3; positioning 78–79; students position themselves, relation 80–81
computer teacher educators 85
computing 1, 2, 4–7, 11–13, 15, 16, 21, 34, 40, 82, 84, 108–110, 118; abstract subject 109; attitudes and beliefs, students 109; careers 3, 4; classes 55, 81, 82; classroom designs 88–90; current position in 11–12; gender imbalance in 109; models for teaching 40–42; pupils, opting out 75–83; research and 77–78
Computing at School (CAS) 5
computing classrooms 35, 54, 84, 85, 88, 91, 93, 103
computing curriculum 5, 7, 9, 12–14, 21, 39, 43, 104, 105
computing education 2, 16, 19–21, 28, 29, 93, 95, 97, 99, 101–103, 118; design and project approaches in 95–105; design-based approaches 97
computing educators 16, 43, 85, 86, 90, 110
computing knowledge 7, 12, 13, 16, 38, 108
computing lessons 5, 54
computing programme 3, 6
computing projects 40, 95, 98, 102–104
computing teachers 1–6, 8, 16, 21, 61, 63, 115, 117, 118
computing teaching: diverse ways to participate 100; encouraging project iteration 100–101; high diversity of creations 100; supportive and authentic audience, creators 101
concept maps 42
constructionism 21, 28
consumers: generation of 114–116
content creator 5
Cook, S. 36
The Coolest Project 96
Cope, B. 25
Criado Perez, C. 52
cross-curricular computing 115
cultural capital 15, 16, 76

curriculum: design 1, 8, 12, 14–16, 117, 118; developing 7–16; revisiting 118
curriculum development: subject and 8–11
Curzon, P. 38

Dalley-Trim, L. 25
Darling-Hammond, L. 105
data 66–67; to adapt planning 90–92; formal quantitative data, teachers 62; in schools, challenges 71–72; for teachers and learners 61–62; in UK schools 60–61
datafication of classroom 66–67
data-informed practice 67–70, 73
design-based approaches 95, 97, 98, 101, 104, 105
design-based learning 100–101
design patterns 97–99
developmental robotics 30
development of pupils' digital skills 93
Develop the Interest of Girls 85, 93
digital default 50–53
digital skills 4, 13, 90, 93, 111, 114; for school leavers 114
diverse technology development workforce 2, 54
Dixon-Roman, E. 53

early career teachers 2–3, 12
educational technologies 61, 62, 66–67, 69, 73
educators 53–55
engagement: multiple means of 35
epistemological pluralism 28–30
evolutionary robotics 30
An Exploration in the Space of Mathematics Education 28–30
Expo 2020 5
expression/action: multiple means of 35

Ferrara, E. 55n2
Fiesler, C. 54
follow-up resources 101
formal informal teaching practices 46, 93
future careers 93, 117

game based learning 104
GCSE Computer Science 11, 12, 75–82
gender 8, 11, 81, 84, 85, 88, 90, 93
gender differences: in computing classrooms 84–93; lesson planning 87–88; pedagogy impact 84–86; relevant role models 86–87
General Certificate of Education 75
General Certificate of Secondary Education (GCSE) 40
General Data Protection Regulation (GDPR) 61
General National Vocational Qualifications (GNVQs) 9
Gershgorn, D. 51
girls 12, 35, 76–78, 80–82, 84–86, 88–90, 109
Google Cloud Vision 48
Google reCAPTCHA Enterprise challenge *48*, 49
Graham, M. 20

"hand coded" algorithms 45
hands-on practices 42
Hidi, S.E. 109
Holland, D. 77

iCub *31*
impact upon subject uptake 85
inclusion 1, 34–36
inclusive classroom 4–5
inclusive computing curriculum 60–73
inclusive digital curriculum: designing 12–14
inclusive learning spaces 84–93
inclusive pedagogies 34–36
inclusive real-world problem 109–110
industry insights: developing your own 117–118
industry perspectives 108–118; interview 109–118
Information and Communication Technology (ICT) 7–16
integrated development environments (IDEs) 63
3 I's (Intent, Implementation and Impact) 3
iSnap 63

Kafai, Y. 100
Kalantzis, M. 25
knowledge, building 14–16
Kozulin, A. 25

Lane, Dale 21
learner-led approaches 36
learning 30–32; frameworks 42; process 19, 22, 24, 28, 41
learning approaches: concrete and abstract 36–37
learning theory: intersections of 67–70
leveling up 11, 39, 75
librarian recommender 22
limited digital skills 66
local communities 110–111
logical thinking 28
Logo 22, 29

machine-learning (ML) 21, 26, 30, 45, 47, 111–113
Machine Learning for Kids (MLfK) 21, 22, 26, 31, 32, 111–113
"Make me Happy" activity 26, *27*
memory 4
Mindstorms 19, 29
Muller, J. 14, 15

National Centre for Computing Education (NCCE) 4, 34
National Curriculum for Computing 3
Noss, R. 20

Ofsted framework 3, 12
Ofsted research report 12
online communities 100
online learning 64–66; platforms 64–66
online scratch community 100–101
online technologies 70, 73
options process 76–79

Pacis, D. 70
pair programming 41, 85, 90, 106
Papert, S. 19–32, 37, 43
pedagogical approaches 2
pedagogical methods 85, 93

personalisation 22
Piaget 21–28
pluralism 29
potential biases in technology development 52, 54, 113
powerful knowledge 14, 15
practical examples 67
practitioners 53–55
PRIMM 41
project-based approaches 43, 95, 101, 102, 105
project-based learning (PBL) 2, 95, 102–106; computing classrooms 103–104; creatively overcoming limitations 104–105
pupil progress 15, 16, 65, 71

questioning techniques 4

Raspberry Pi Foundation 5
Reaume, A. H. 47, 49
representation: multiple means of 35
Resnick, M. 39, 43, 98, 104
Roehampton Annual Computing Education Report 11
Rogoff, B. 96
Roque, R.V. 101
Rusk, N. 43, 104

school curriculum 15
school leavers 10, 114, 115
Scottish Alexa 55n1
semantic profiles 37–38
semantic waves 38
senior leader perspectives 61
"Shutdown or Restart" report 10, 21
skills for industry 13, 93, 115
social media influencer 5
social media profile 5, 49, 51, 73, 115
special educational needs and disabilities (SEND) 3, 34, 35, 61
STEM Ambassadors 6
structured observations 42

student–teacher lesson scenarios **91–92**
Sumpter, D. 50
support careers 6, 117
support computing teachers 1, 11

teachers 81–83
teaching: concrete and abstract in 38–39
teaching computing 4, 28, 36, 40, 41, 90, 118
teaching technique: examples and design patterns 98–99; iterative design 97–98; PBL, computing classrooms 103–104; use-modify-create model 99–100
Technical and Vocational Education Initiative (TVEI) 9
"Ten quick tips for teaching programming" 64
Thought and Language 24
time challenges 104
trainee teachers 2–3
Turkle, S. 24, 28, 37

Universal Design for Learning (UDL) 34, 35–36
unplugged activities 41
unstructured observations 42

verbal protocols 42
video conferencing tools 65
Vygotsky 21–28

Weegar, M.A. 70
Werner, L. L. 85
Whittaker, M. 52
Whitty, G. 20
widen participation 96, 100
Wolf, M. 51
workplace: schools preparing young people 116–117

Young, M. 14, 15
youth culture 110–111

Zook, M. A. 20

For Product Safety Concerns and Information please contact our EU
representative GPSR@taylorandfrancis.com
Taylor & Francis Verlag GmbH, Kaufingerstraße 24, 80331 München, Germany

www.ingramcontent.com/pod-product-compliance
Lightning Source LLC
Chambersburg PA
CBHW081025240426
43661CB00074B/2859